MW00851134

This book provides an introducti<del>on</del> <del>which un</del>derlie the modern understanding o<del>f</del> physical systems which exhibit the property of scale invariance self-similarity. This is most clearly illustrated in materials, such as magnets or fluids, in the vicinity of a second order phase transition. The theoretical framework for understanding these phenomena, known as the renormalization group, first arose in the late 1960s and has evolved into a common language used by workers in such diverse fields as particle physics, cosmology, neural networks and biophysics, as well as the more conventional aspects of condensed matter physics.

Beginning with a brief review of phase transitions in simple systems and of mean field theory, the text then goes on to introduce the core ideas of the renormalization group. Following chapters cover phase diagrams, fixed points, cross-over behaviour, finite-size scaling, perturbative renormalization methods, low-dimensional systems, surface critical behaviour, random systems, percolation, polymer statistics, critical dynamics and conformal symmetry. The book closes with an appendix on Gaussian integration, a selected bibliography, and a detailed index. Many problems are included.

The emphasis throughout is on providing an elementary and intuitive approach. In particular, the perturbative method introduced leads, among other applications, to a simple derivation of the epsilon expansion in which all the actual calculations (at least to lowest order) reduce to simple counting, avoiding the need for Feynman diagrams.

CAMBRIDGE LECTURE NOTES IN PHYSICS 5
General Editors: P. Goddard, J. Yeomans

Scaling and Renormalization
in Statistical Physics

CAMBRIDGE LECTURE NOTES IN PHYSICS
1. Clarke: The Analysis of Space-Time Singularities
2. Dorey: Exact S-Matrices in Two Dimensional Quantum
Field Theory
3. Sciama: Modern Cosmology and the Dark Matter Problem
4. Veltman: Diagrammatica–The Path to Feynman Rules
5. Cardy: Scaling and Renormalization in Statistical Physics

Scaling and Renormalization in Statistical Physics

JOHN CARDY

University of Oxford

CAMBRIDGE
UNIVERSITY PRESS

PUBLISHED BY THE PRESS SYNDICATE OF THE UNIVERSITY OF CAMBRIDGE
The Pitt Building, Trumpington Street, Cambridge, United Kingdom

CAMBRIDGE UNIVERSITY PRESS
The Edinburgh Building, Cambridge CB2 2RU, UK
40 West 20th Street, New York, NY 10011–4211, USA
10 Stamford Road, Oakleigh, VIC 3166, Australia
Ruiz de Alarcón 13, 28014 Madrid, Spain
Dock House, The Waterfront, Cape Town 8001, South Africa

http://www.cambridge.org

© Cambridge University Press 1996

This book is in copyright. Subject to statutory exception
and to the provisions of relevant collective licensing agreements,
no reproduction of any part may take place without
the written permission of Cambridge University Press.

First published 1996
Reprinted (with corrections) 1997, 2000

Printed in the United Kingdom at the University Press, Cambridge

Typeset by the author

A catalogue record for this book is available from the British Library

ISBN 0 521 49959 3 paperback

To the memory of my father

Contents

	Preface	*page* xiii
1	**Phase transitions in simple systems**	**1**
1.1	Phase diagrams	5
1.2	Simple models	10
	Exercises	14
2	**Mean field theory**	**16**
2.1	The mean field free energy	16
2.2	Critical exponents	20
2.3	Mean field theory for the correlation function	22
2.4	Corrections to mean field theory	24
	Exercises	26
3	**The renormalization group idea**	**28**
3.1	Block spin transformations	30
3.2	One-dimensional Ising model	34
3.3	General theory	40
3.4	Scaling behaviour of the free energy	43
3.5	Critical exponents	45
3.6	Irrelevant eigenvalues	48
3.7	Scaling for the correlation functions	49
3.8	Scaling operators and scaling dimensions	52
3.9	Critical amplitudes	55
	Exercises	59
4	**Phase diagrams and fixed points**	**61**
4.1	Ising model with vacancies	61
4.2	Cross-over behaviour	67
4.3	Cross-over to long range behaviour	71
4.4	Finite-size scaling	72
4.5	Quantum critical behaviour	76

Exercises   81

5   **The perturbative renormalization group**   83
5.1   The operator product expansion   84
5.2   The perturbative renormalization group   86
5.3   The Ising model near four dimensions   90
5.4   The Gaussian fixed point   92
5.5   The Wilson–Fisher fixed point   94
5.6   Logarithmic corrections in $d = 4$ 103
5.7   The $O(n)$ model near four dimensions   104
5.8   Cubic symmetry breaking   107
    Exercises   109

6   **Low dimensional systems**   111
6.1   The lower critical dimension   111
6.2   The two-dimensional XY model   113
6.3   The solid-on-solid model   117
6.4   Renormalization group analysis   120
6.5   The $O(n)$ model in $2 + \epsilon$ dimensions   127
    Exercises   131

7   **Surface critical behaviour**   133
7.1   Mean field theory   134
7.2   The extraordinary and special transitions   136
7.3   Renormalization group approach   138
    Exercises   144

8   **Random systems**   145
8.1   Quenched and annealed disorder   145
8.2   The Harris criterion   148
8.3   Perturbative approach to the random fixed point   150
8.4   Percolation   153
8.5   Random fields   161
    Exercises   167

9   **Polymer statistics**   169
9.1   Random walk model   169
9.2   The Edwards model and the Flory formula   170
9.3   Mapping to the $O(n)$ model   172
9.4   Finite concentration   177
9.5   Other applications   178

Exercises 181

10 Critical dynamics 183
10.1 Continuum models 186
10.2 Discrete models 189
10.3 Dynamic scaling 192
10.4 Response functional formalism 193
10.5 Other dynamic universality classes 196
10.6 Directed percolation 200
Exercises 204

11 Conformal symmetry 206
11.1 Conformal transformations 209
11.2 Simple consequences of conformal symmetry 211
11.3 The stress tensor 215
11.4 Further developments 222
11.5 The c-theorem 223
Exercises 225

Appendix: Gaussian integration 227
Selected bibliography 229
Index 234

Preface

Scaling concepts play a central role in the analysis of the ever more complex systems which nowadays are the focus of much attention in the physical sciences. Whether these problems relate to the very large scale structure of the universe, to the complicated forms of everyday macroscopic objects, or to the behaviour of the interactions between the fundamental constituents of matter at very short distances, they all have the common feature of possessing large numbers of degrees of freedom which interact with each other in a complicated and highly non-linear fashion, often according to laws which are only poorly understood. Yet it is often possible to make progress in understanding such problems by isolating a few relevant variables which characterise the behaviour of these systems on a particular length or time scale, and postulating simple *scaling relations* between them. These may serve to unify sets of experimental and numerical data taken under widely differing conditions, a phenomenon called *universality*. When there is only a single independent variable, these relations often take the form of power laws, with exponents which do not appear to be simple rational numbers yet are, once again, universal.

The existence of such scaling behaviour may often be explained through a framework of theoretical ideas loosely grouped under the term *renormalization*. Roughly speaking, this describes how the parameters specifying the system must be adjusted, under putative changes of the underlying dynamics, in such a way as not to modify the measurable properties on the length or time scales of interest. The simple postulate of the existence of a fixed point of these renormalization flows is then sufficient to explain qualitatively the appearance of universal scaling laws. Unfortunately, for most examples of complex systems, such a renormalization approach has not, as yet, been put on a systematic basis starting from the underlying microscopic dynamics. In trying to understand scaling arguments applied to such problems it is often difficult, especially for newcomers, to understand why certain variables should be neglected while others are retained in such scaling descriptions, and why in some cases power law relations should hold while they fail in others.

Fortunately, there is a class of physical problems within which the concepts of scaling and renormalization may be derived systematically, and which therefore have become a paradigm for the whole approach. These concern equilibrium critical behaviour. The systems which exhibit such behaviour are governed by the simple and well understood laws of statistical mechanics. Indeed, along with the high energy behaviour of quantum field theories, this was the area of physics in which the concepts of renormalization were first formulated. Although the subject of equilibrium critical behaviour is, apart from a few unsolved problems, no longer of the greatest topical theoretical or experimental interest, its study is nonetheless important in providing a solid grounding to anyone who wishes to go on to attempt to understand scaling and renormalization in more esoteric systems. Yet the typical student in condensed matter theory faces a problem in trying to accomplish this. Historically, the subjects of renormalization in quantum field theory (as applied to particle physics) and in equilibrium critical behaviour have developed in parallel. This is no coincidence – the two sets of problems have, mathematically, a great deal in common, and, indeed, the most systematic formulation of the subject relies heavily on the property of renormalizability in quantum field theory. However, much of the qualitative structure of renormalization may be introduced through the alternative real space methods which are both simple and appealing. But students who learn this approach, and then wish to go further in existing accounts of the subject, must make a complete change of gears to momentum space methods which require a great deal of investment of time and effort in digesting the whole formalism of Feynman diagrams and renormalization theory. As a result, the study of the subject rapidly becomes overladen with formalism, and the student, if he or she is lucky, has just about time to learn how to calculate the critical exponents of the Ising model in $4 - \epsilon$ dimensions before the course comes to an end. The average student thereby often misses out on any account of the tremendously wide range of problems, even within critical behaviour, on which these methods may be brought to bear.

In my opinion, these field theoretic details are appropriate only for the relatively small number of students who wish to go on and apply these methods to particle physics, or for those who really need to compute critical exponents to $O(\epsilon^2)$ and higher. For the majority, whose goal is to understand how scaling and renormalization ideas might be applied to the rich variety of complex phenomena apparent in many other branches of the physical sciences, the main object is to learn the concepts, and the best way to do this is by covering as many examples as possible. This small book was written with this goal in mind. It is, in fact, based on a set of lectures which were given, in various incarnations, to physics graduate students at Santa Barbara and Oxford. A significant fraction of the audience consisted of students planning to do experimental rather than theoretical research.

I have assumed that the reader has already had a basic course in statistical mechanics, and, indeed, has had some exposure to critical phenomena, a subject which is, nowadays, often discussed in such courses. However, for completeness, the basic phenomena and some simple models are recalled in the first chapter. Next comes a discussion of the 'classical' approach to critical behaviour through mean field theory, before the renormalization group idea is introduced. As mentioned above, the simplest conceptual route to renormalization concepts is through real space methods, and I have chosen this approach. At this level, all of the qualitative properties of scaling and universality may then be discussed.

However, it is also important that the student understand how quantitative methods, such as the ϵ-expansion, come about. In this book I describe an approach, which is certainly not new but deserves to become better known, by which at least the first order perturbative renormalization group equations may be derived from a simple continuum real space approach, thus linking up directly with the earlier more intuitive considerations. It relies on the operator product expansion of field theory, an impressive sounding name for something which is basically very intuitive and simple, and easy to calculate with in lowest order. As a result, the $O(\epsilon)$ results for the critical exponents emerge as a consequence of elementary combinatorics, with no Feynman diagrams required at all! This approach also lays stress on the modern idea of scaling 'operators' and their associated scaling dimensions as being the central objects of attention, rather than derived quantities like the traditionally defined critical exponents.

Although I have deliberately tried to avoid couching the discussion in the language of quantum field theory, a few of its elementary results, particularly those of Gaussian integration and the combinatorial version of Wick's theorem, are nonetheless required. The details of these are summarised in a brief Appendix, for readers unfamiliar with these simple formulae.

After this, the book embarks on a tour of many of the important applications of renormalization group to critical phenomena. After a few simple generalisations of the Ising magnet in $4 - \epsilon$ dimensions, we descend to the neighbourhood of two dimensions and show how the same perturbative renormalization group methods which gave us the ϵ-expansion may be applied to the famous examples of the XY model and other systems with continuous symmetries. Then come accounts of the application of similar methods to critical behaviour near surfaces, to systems with quenched random impurities, and to the configurational statistics of large polymers in solution. These are all problems in equilibrium critical behaviour, but the next chapter brings in the dynamics. This is a tremendously rich subject, indeed one which deserves a whole book in itself at this level, and it is therefore impossible to do it justice in a single chapter. However, I have tried to include a number of examples apart from the standard ones, including in particular directed percolation, an example from the rapidly expanding subject of dynamic critical behaviour in systems far from equilibrium. Finally, the tour ends with an elementary account of some of the recent developments in the application of the ideas of conformal symmetry to equilibrium critical behaviour. The non-mathematical reader may find this section slightly harder going than the earlier chapters, although all that is in fact required is a basic knowledge of tensor calculus and complex analysis.

Unfortunately, many important examples of scaling in statistical physics have been omitted in this survey, due to reasons of lack of space and/or expertise on the part of the author. In particular, I would have liked to have spent time on the problems of fluctuating interfaces and of spin glasses, where the concepts of scaling are amply illustrated, but these subjects are too complicated for inclusion in a single course. Similarly, the modern approach to localisation of waves and electrons in random systems is replete with scaling arguments, but it too requires too extensive an introduction. The dynamics of phase ordering following a rapid temperature quench is another fascinating related subject where scaling arguments play a central role, but for which, as yet, no systematic renormalization approach has been formulated.

A more profound apology is required for the lack of any detailed reference to comparison with experimental data. I hope that this does not create the wrong impression. The subject of critical phenomena is one which is, ultimately, driven by observation and experiment, and it is important that all theorists continue to bear this in mind. However, the basic experiments which established the phenomena of scaling and universality in critical behaviour were performed some time ago, and their results are by now adequately summarised in a number of standard references. It is not the purpose of this book to make detailed comparison with experimental results on particular systems, but rather to emphasise the generality of the principles involved. In this sense the current status of the theory is akin to that of quantum mechanics, where, in similarly introductory texts, it is considered adequate nowadays to illustrate the theoretical principles with applications to simple and rather idealised systems, rather than by comparison with detailed experimental data.

Since this is an introductory account, I have not included bibliographic references in the text. Rather, I have provided a list of selected sources of further reading at the end. I have also included a number of exercises, the aim of which is to lead the inquisitive reader into further examples and extensions of the ideas discussed in the text.

I thank my graduate students and colleagues at Santa Barbara and Oxford who have helped me formulate the material of this book over the years. I am particularly grateful to Benjamin Lee for a careful reading of the manuscript, and to Reinhard Noack for helping produce the Ising model pictures in Chapter 3.

1

Phase transitions in simple systems

Take a large piece of material and measure some of its macroscopic properties, for example its density, compressibility or magnetisation. Now divide it into two roughly equal halves, keeping the external variables like pressure and temperature the same. The macroscopic properties of each piece will then be the same as those of the whole. The same holds true if the process is repeated. But eventually, after many iterations, something different must happen, because we know that matter is made up of atoms and molecules whose individual properties are quite different from those of the matter which they constitute. The length scale at which the overall properties of the pieces begin to differ markedly from those of the original gives a measure of what is termed the *correlation length* of the material. It is the distance over which the fluctuations of the microscopic degrees of freedom (the positions of the atoms and suchlike) are significantly correlated with each other. The fluctuations in two parts of the material much further apart than the correlation length are effectively disconnected from each other. Therefore it makes no appreciable difference to the macroscopic properties if the connection is completely severed.

Usually the correlation length is of the order of a few interatomic spacings. This means that we may consider really quite small collections of atoms to get a very good idea of the macroscopic behaviour of the material. (This statement needs qualification. In reality, small clusters of atoms will exhibit very strong surface effects which may be quite different from, and dominate, the bulk behaviour. However, since this is only a thought experiment, we may imagine employing the theoretician's device of periodic boundary conditions, thereby eliminating them.) However, the actual value of the correlation length depends on the external conditions determining the state of the system, such as the temperature and pressure. It is well known that systems may abruptly change their macroscopic behaviour as these quantities are smoothly varied. The points at which this happens are called critical points, and they usually mark a phase transition from one state of matter to another. There are basically two possible ways in which such a transition may occur. In the first scenario, the two (or more) states on either side of the critical point also co-exist exactly at the critical point. However, even then they are distinct from each other, in that they have different macroscopic properties. Slightly away from the critical point, however, there is generically a unique phase whose properties are continuously connected to one of the co-existent phases at the critical point. In that case, we should expect to find discontinuous behaviour in various thermodynamic quantities as we pass through the critical point, and therefore from one stable phase to another. Such transitions are termed *discontinuous* or *first-order*. Well-known examples are the melting of a three-dimensional solid, or the condensation of a gas into a liquid. In fact, such transitions often exhibit hysteresis, or memory effects, since the continuation of a given state into the opposite phase may be metastable so that the system may take a macroscopically long time to readjust. The correlation length at such a first-order transition is generally finite.

However, the situation is quite different at a *continuous* transition, where the correlation length becomes effectively infinite. The fluctuations are then correlated over all distance scales, which thereby forces the whole system to be in a unique, critical, phase. At a continuous transition, therefore, the two (or more) phases on either side of the critical point must become identical as it is approached. Not only does the correlation length diverge in a continuous fashion as such a critical point is approached, but the differences in the various thermodynamic quantities between the competing phases, like the energy density and the magnetisation, go to zero smoothly. It is the task of the theory to explain this behaviour in a quantitative manner. Simple examples of continuous transitions, to be described in more detail below, occur at the Curie temperature in a ferromagnet, and at the liquid–gas critical point in a fluid.

The fact that a very large number of degrees of freedom are strongly correlated with each other makes the study of continuous phase transitions intrinsically difficult. By their nature, these phenomena are not amenable to normal perturbative methods. It is only within the last twenty-five years or so that analytic methods have been developed for dealing with such problems. These methods constitute a whole new way of thinking about such phenomena, which is called the *renormalization group*.

Although systems with large correlation lengths might appear to be very complex, they also exhibit some beautiful simplifications. One of these is the phenomenon of universality. Many properties of a system close to a continuous phase transition turn out to be largely independent of the microscopic details of the interactions between the individual atoms and molecules. Instead, they fall into one of a relatively small number of different classes, each characterised only by global features such as the symmetries of the underlying hamiltonian, the number of spatial dimensions of the system, and so on. This phenomenon finds a simple and natural explanation within the framework of the renormalization group. Typically, close to a critical point, the correlation length and the other thermodynamic quantities exhibit power-law dependences on the parameters specifying the distance away from the critical point. These powers, or *critical exponents*, are pure numbers, usually not integers or even simple rational numbers, which depend only on the universality class. One of the basic theoretical challenges, then, is to explain why such non-trivial powers should occur, and to predict their actual values.

The occurrence of power behaviour in the laws describing a system is a symptom of *scaling* behaviour. Such dependences occur, of course, in problems at many levels in the physical sciences. In the most elementary cases, they are simply the result of dimensional analysis. For example, once the inverse square law $1/r^2$ for the gravitational acceleration of a body a distance r from a point mass is assumed, Kepler's law that the orbital period $T \propto r^{3/2}$, follows immediately. (Note that, in general, such elementary arguments lead to simple rational numbers for the exponents.) However, most physical problems present us with more than one length scale, and simple dimensional analysis is no longer adequate. Physical quantities may then depend in an arbitrarily complicated manner on the dimensionless ratios of these scales. Nevertheless, simplifications may occur if there is a wide *separation of scales* in the problem. In that case, it may be permissible to neglect the shorter length scales when discussing the large scale physics. This, in fact, already occurs in the problem of planetary motion, where it may be shown that the finite radius of the bodies has no significant effect.

Similar results follow in the so-called 'classical' approaches to the problem of continuous phase transitions. The approximate theory to be developed in Chapter 2 implies that, close to the critical point of a ferromagnet, the correlation function $G(r)$ of the local fluctuations in the magnetisation obeys Laplace's equation over distance scales r much less than the correlation length ξ, and therefore exhibits a $1/r$ behaviour in three dimensions, like the gravitational potential. This problem would appear to have two potentially important length scales: the microscopic length a which specifies the typical distance between the fluctuating magnetic degrees of freedom, and the correlation length ξ. Thus, we might be tempted to argue, in analogy with the planetary problem, that when $\xi \gg a$, that is, very close to the critical point, the microscopic length a may be ignored. Under that assumption, dimensional analysis then implies that G has the form

$$G(r) = \frac{1}{r} f(r/\xi), \qquad (1.1)$$

where f is some function, as yet unknown. From the magnetic correlation function, we may infer the susceptibility $\chi \propto \int G(r) d^3 r$. Substituting in the above form, we then find, after a change of integration variables, that $\chi \propto \xi^2$. This simple power law is essentially a consequence of dimensional analysis, and is a typical result of the 'classical' approach to critical behaviour, which predates the renormalization group.

However, both experiments and studies of simplified lattice models indicate that the above result is incorrect. The reason is that, unlike the Kepler problem, critical behaviour is dominated by the effects of fluctuations, and these fluctuations take place on all length scales, all the way down to the microscopic distance a. It is therefore not permissible simply to neglect a, even when it is much smaller than the characteristic length ξ. It might then appear that little more can be said without further detailed analysis, since (1.1) is now replaced by the weaker relation

$$G(r) = \frac{1}{r} f(r/\xi, a/\xi). \tag{1.2}$$

However, it turns out that, for $a/\xi \ll 1$, although the function f is not independent of this ratio, it nevertheless exhibits a simple power law dependence, proportional to $(a/\xi)^\eta$, where η is a small but non-zero exponent. As a consequence, the dependence of the susceptibility on the correlation length has the form

$$\chi \propto a^\eta \, \xi^{2-\eta}. \tag{1.3}$$

Thus, scaling is, in a sense, recovered, but with an exponent $2-\eta$ which does not follow from straightforward dimensional analysis. The deviation η is an example of an *anomalous dimension*. The existence of such behaviour demands, of course, an explanation. It will turn out that this arises quite naturally within the framework of the renormalization group.

Before discussing these general properties further, however, it is useful to have at hand some very simple physical examples of continuous critical behaviour in which these concepts may be illustrated in a concrete fashion.

1.1 Phase diagrams

Uniaxial ferromagnets

In a ferromagnet there are two interesting external parameters which may be varied: the temperature T and the applied magnetic field H. In the most straightforward case, the local magnetisation is constrained to lie parallel or anti-parallel to a particular axis. The phase diagram (Figure 1.1) is simple. All the thermodynamic quantities (e.g. specific heat, susceptibility) are smooth analytic functions of T and H except on the line $H = 0, T \leq T_c$. Across the line $T < T_c$, the magnetisation M, as a function of H, is discontinuous, having the form illustrated in Figure 1.2a. This discontinuity is characteristic of a first-order transition, with a finite correlation length. As T approaches the Curie point T_c, from below, however, the discontinuity approaches zero, and the correlation length at the transition diverges. The point $H = 0, T = T_c$ is an example of a *critical end point*, at which the first-order transition becomes continuous.

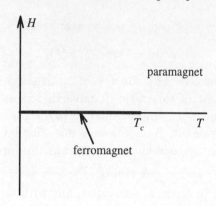

Figure 1.1. Phase diagram of a uniaxial ferromagnet.

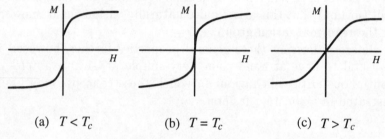

(a) $T < T_c$ (b) $T = T_c$ (c) $T > T_c$

Figure 1.2. Magnetisation versus applied field, for various temperatures.

When $T < T_c$, the two limits $H \to 0+$ and $H \to 0-$ give different possible values $\pm M_0$ for the magnetisation. Which one the system chooses depends on its previous history. This is an example of *spontaneous symmetry breaking*: although the hamiltonian is invariant under simultaneous reversal of all the local magnetic degrees of freedom, this symmetry is not respected by the equilibrium thermodynamic state. The onset of such symmetry breaking is a common, although not a universal, characteristic of continuous critical points. The magnetisation M, whose value measures the amount of magnetic order in the material, is called the *order parameter* for this transition. In most, but not all, examples of critical behaviour, it is possible to identify one or more such order parameters, and the behaviour of their local fluctuations often provides a useful way of characterising the nature of the transition.

As mentioned above, most of the quantities of interest exhibit power law behaviour sufficiently close to the critical point. We now give the definitions of the principal critical exponents which characterise these power laws for the case of a ferromagnet. It is useful to define two dimensionless measures of the deviation from the critical point: the *reduced temperature* $t \equiv (T - T_c)/T_c$, and the reduced external magnetic field $h = H/k_B T_c$. The exponents are

α: The specific heat in zero field $C \sim A|t|^{-\alpha}$, apart from terms which are regular in t. Note that, in principle, one should consider the possibility of different exponents α and α' for $t > 0$ and $t < 0$ respectively. However, it is an immediate consequence of the renormalization group (in agreement with other exact results) that $\alpha = \alpha'$, and we shall henceforth cease to make a distinction. α can be positive or negative, corresponding to either a divergent spike or a cusp in the specific heat when plotted against T. Although the exponent α is universal, the *amplitude* A is not, and, moreover, $A \neq A'$ in general. However, a prediction of the renormalization group is that the *ratio* A'/A is universal (see Section 3.9).

β: The spontaneous magnetisation $\lim_{H \to 0+} M \propto (-t)^\beta$.

γ: Zero field susceptibility $\chi \equiv (\partial M/\partial H)|_{H=0} \propto |t|^{-\gamma}$. Once again, in principle different exponents should be defined for different signs of t, but in fact theory indicates that they should be equal.

δ: At $T = T_c$, the magnetisation varies with h according to $M \propto |h|^{1/\delta}$.

ν: A more quantitative measure of the correlation length ξ than given above is through the asymptotic behaviour $G(r) \propto e^{-r/\xi}/r^{(d-1)/2}$ (for $r \gg \xi$) of the correlation function of the fluctuations in the local magnetisation. Alternatively, it may be defined through the second moment of this quantity, $\xi^2 = \sum_r r^2 G(r) / \sum_r G(r)$. In either case, it diverges as $t \to 0$, with $h = 0$, according to $\xi \propto |t|^{-\nu}$. Once again, this has a meaning either side of the critical point.

η: Exactly at the critical point, the correlation function does not decay exponentially, but rather according to $G(r) \propto 1/r^{d-2+\eta}$.

z: Finally, there is an exponent relating to the time-dependent properties close to the critical point. For example, the typical relaxation time τ diverges, as the critical point is approached, according to $\tau \propto \xi^z$. Since this exponent does not relate to the static equilibrium properties, however, further discussion will be deferred until Chapter 10.

This completes the commonly observed zoo of exponents for this critical point. Although this nomenclature has come to be accepted for historical reasons, it will become clear in Chapter 3 that these critical exponents are not the most fundamental quantities from the theoretical perspective. Rather, they are simply derived from a smaller set of numbers called the scaling dimensions.

Many of the definitions given above may be taken over without modification to the Curie point in more general ferromagnets, for example Heisenberg magnets where the local moments are free to rotate in three dimensions. The only difference is in the low-temperature phase, when a distinction must be drawn between fluctuations of the local magnetisation parallel to, and perpendicular to, the direction of the spontaneous magnetisation. In this case, the transverse susceptibility, the response of the magnetisation to an applied field perpendicular to the spontaneous magnetisation, remains infinite throughout the low-temperature phase, and the critical exponent γ must then be defined in terms of the longitudinal susceptibility only. Similar generalisations apply to the vicinity of the Néel point in antiferromagnets. In this case, it is the sublattice, or staggered, magnetisation which plays the role of the order parameter.

Simple fluids

The phase diagram of a generic substance in the temperature-pressure plane usually has the form shown in Figure 1.3. Let us focus on the part of the phase diagram close to the liquid–gas critical point at (T_c, p_c). It looks quite similar to that for the ferromagnet in the (T, H)–plane. Across the liquid–gas phase boundary, the density ρ is discontinuous. The jump in the density $\rho_{\text{liquid}} - \rho_{\text{gas}}$ approaches zero at the critical end point (T_c, p_c). The isotherms, curves of p *versus* ρ at constant T, are illustrated in Figure 1.4.

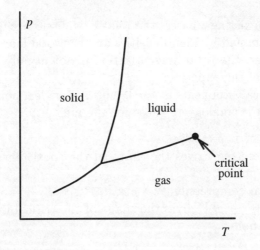

Figure 1.3. Phase diagram of a typical substance.

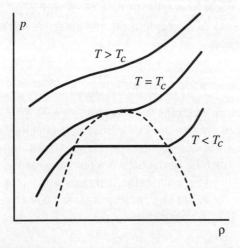

Figure 1.4. Isotherms near the liquid–gas critical point. The dashed line is the coexistence curve.

Comparing this with the graphs of the magnetisation in a ferromagnet shown in Figure 1.2, we see that $p-p_c$ is analogous to the applied field H, and $\rho-\rho_c$ to the magnetisation M. The main physical difference between the systems is that, in a fluid the overall average density is usually fixed. This means that, underneath the coexistence curve shown in Figure 1.4, the system separates into two coexisting phases, the relative volume occupied by each being determined by their densities and the total mass. For a magnet, however, the total magnetisation is not usually fixed by the conditions of the experiment.

The critical exponents at the liquid–gas critical point are therefore defined in analogy to those for the magnet:

- $C_V \propto |t|^{-\alpha}$ at $\rho = \rho_c$;
- $\rho_L - \rho_G \propto (-t)^\beta$ gives the shape of the coexistence curve near the critical point;
- isothermal compressibility $\chi_T \propto |t|^{-\gamma}$;
- $p - p_c \propto |\rho_L - \rho_G|^\delta$ gives the shape of the critical isotherm near the critical point.

The exponents ν and η are defined as for the ferromagnet, with $G(r)$ now being the density-density correlation function.

One of the most remarkable results of universality is that the critical exponents of a simple fluid are identical with those of uniaxial ferromagnets.

1.2 Simple models

These equilibrium systems with many degrees of freedom, are, of course, governed by the laws of statistical mechanics. This means that the mathematical problem is at least well-defined. Physical quantities are given in terms of averages with respect to the Gibbs distribution $e^{-\beta\mathcal{H}}$.† In particular, thermodynamic quantities such as the specific heat or the magnetisation are given by suitable derivatives of the partition function

$$Z = \mathrm{Tr}\, e^{-\beta\mathcal{H}}, \qquad (1.4)$$

or, equivalently, of the free energy $F = -\beta^{-1} \ln Z$. Even when the hamiltonian $\mathcal{H}$ is relatively simple, computing Z is usually very difficult. For a realistic hamiltonian, it is a hopeless task. This has led to the consideration of drastically simplified models which, it is hoped, nonetheless capture the essential physics. However, there is a unique advantage in approaching these kinds of problem through the study of models, which is not present in most areas of physics.

---

† There should be no confusion between $\beta \equiv (k_B T)^{-1}$ and the critical exponent denoted by the same symbol.

In the study of atomic phenomena, for example, one tries to invent a model which is as close to the true hamiltonian as possible, but is nevertheless solvable, exactly or approximately. Once the parameters have been suitably adjusted, the results of such studies are then expected to match the observed phenomena closely, but, not of course exactly, since any model inevitably omits some effects. Once rough agreement has been obtained, the model may then be refined, so as to improve the accuracy of the description.

However, because of the phenomenon of universality in critical phenomena, one may hope to obtain results from very simple models which *exactly* match the behaviour of real systems, at least in those aspects which are universal. Moreover, in general, these universal features should be independent of the microscopic parameters of the model, and there is therefore no question of parameter-fitting: the theory is either right or wrong. In reality the situation is not quite as simple as this. Real or numerical experiments never probe the true asymptotic region close to the critical point, and therefore so-called corrections to scaling, which contain non-universal and therefore adjustable parameters, often need to be incorporated to obtain a good fit. Nevertheless, it is true that theoretical models of critical behaviour may be tested with much greater precision than occurs in most other areas of condensed matter physics.

Even so, only a few of the simplest models in low dimensions may be solved exactly, and although the mathematical methods involved constitute a fascinating subject in their own right, the details of this analysis have little bearing on the kinds of question the physicist would like to have answered. This is partly because the exact solution of a model does not, in general, differentiate between those results which are universal and those which are not. It is then by no means clear which properties of the exact solution of the simplified model are supposed to carry over to real systems. However, the renormalization group approach does make such a distinction, through the arguments we shall develop in Chapter 3. From this point of view, then, model hamiltonians are more usefully regarded as frameworks within which to describe the important microscopic features of a given universality class, rather than as objects for exact solution.

An immediate simplification which may be made in almost all

systems comes from the observation that, at finite temperatures, the critical thermal fluctuations completely dominate the quantum mechanical ones. This statement will be made more quantitative in Section 4.5. Thus, the simplified models we study are almost always described by a *classical* hamiltonian $\mathcal{H}$, and the calculation of the partition function involves a sum over classical phase space. This is true even though the underlying physics driving the phenomena (e.g. magnetic exchange interactions, superfluidity) may be intrinsically quantum mechanical in nature.

Magnets

The simplest way of modelling a ferromagnet is to imagine quantum mechanical spins $s(r)$ localised on the sites r of a lattice. The spins interact in pairs with a hamiltonian $\mathcal{H} = -\frac{1}{2}\sum_{r,r'} J(r,r')$ $s(r)\cdot s(r')$, where $J(r,r')$ falls off with the distance between r and r'. If $J > 0$ the interaction is ferromagnetic. Application of an external magnetic field corresponds to the addition of a term $-\mu\mathbf{H}\cdot\sum_r s(r)$. This model is the quantum Heisenberg model. For the reasons mentioned above, however, at finite temperature we may ignore the quantum aspect of the spins, and regard the $s(r)$ as classical objects whose configuration space corresponds to the points on the surface of a sphere.

In a real system the spins usually lie in a crystalline lattice which does not possess full rotational symmetry. There may then be crystalline fields acting on the spins which make them prefer to align along certain axes. In the case of uniaxial anisotropy, the spins prefer to lie along, for example, the z-axis, and we may then restrict the allowed values of the degrees of freedom to be $s^z(r) = \pm 1$, in suitable units. This gives the *Ising model*, perhaps the most famous and fundamental model of critical behaviour. Alternatively, the spins may prefer to lie down in the xy–plane, so that $s(r)$ has only two components, satisfying $s^x(r)^2 + s^y(r)^2 = 1$. This gives the XY or *planar model*.

The XY model has another important realisation. In superfluid helium, a finite fraction of the degrees of freedom of the system condense into a macroscopic quantum state, whose wave function $\Psi(r)$ is a complex number. It turns out that for most purposes we may ignore its quantum mechanical origin and treat Ψ as a classical field. Moreover, it is the fluctuations in the phase of Ψ, rather than in its modulus, which dominate the transition into the normal state. Thus we may regard the length of Ψ as fixed. Moreover, the short-range repulsion between the helium atoms may be modelled by localising the degrees of freedom to lie on a lattice, thereby arriving at a model mathematically identical to the XY model. Similar considerations apply to superconductors, which also provide examples of the XY universality class.

More generally, we may consider a spin having n components. This gives the n-vector or $O(n)$ model. The above three cases correspond to $n = 3, 1, 2$ respectively, but it is mathematically possible to think about this model for arbitrary n. As we shall see in Chapter 9, even the limit $n \to 0$ has a physical interpretation in describing the statistics of long polymer chains.

Fluids

A simple classical fluid may be specified by the positions $r_1, \ldots, r_N$ of its N particles. It is often convenient to work in the grand canonical ensemble, in which case the grand partition function is

$$\Xi = \sum_N \frac{\zeta^N}{N!} \int e^{-\beta \sum_{i<j} V(r_i - r_j)} \, d^3 r_1 \ldots d^3 r_N, \qquad (1.5)$$

where ζ is proportional to the fugacity, and we consider only a two-body interaction V, which is supposed to be short-ranged and attractive, with a hard core repulsive component. This model is already quite difficult to analyse using renormalization group methods, and it is common to replace it with an idealisation called the *lattice gas* model. The particles are now assumed to occupy the sites r of a regular lattice, and the hard core interaction is modelled by restricting the occupation number $n(r)$ to take only the values 0 or 1. The attractive part of the potential is described by the interaction $-2\sum_{r,r'} J(r, r') n(r) n(r')$, where $J > 0$. Then

$$\Xi = \sum_{n(r)=0,1} \zeta^{\sum_r n(r)} e^{2\beta \sum_{r,r'} J(r,r') n(r) n(r')}. \qquad (1.6)$$

This may be cast in a more familiar form by defining $s(r) = 2n(r) - 1$, which takes the values ± 1. The terms in the exponential in (1.6) are then, apart from an unimportant constant,

$$\tfrac{1}{2}\beta \sum_{r,r'} J(r,r')s(r)s(r') + \beta H \sum_r s(r), \qquad (1.7)$$

where $H = \tfrac{1}{2}k_B T \ln \zeta + \sum_{r'} J(r,r')$. This is the hamiltonian for the Ising model in a magnetic field! Thus, as long as the various rather crude approximations which led to this result are justified on the grounds of universality, the simple liquid–gas critical point and the Curie point of uniaxial ferromagnets should be in the same universality class, described by the simple Ising model. It is interesting to note that the hamiltonian of the Ising model in zero field has a symmetry under simultaneous reversal of all the spins $s(r) \rightarrow -s(r)$. It is this symmetry which is broken spontaneously below the critical temperature. However, the simple fluid has, in general, no such symmetry in its microscopic dynamics. It is only close to the critical point that this symmetry emerges. For example, the isotherms in Figure 1.4 are symmetrical under reflection about the point (ρ_c, p_c) only close to this point. This is an example of a common property of critical systems, that their symmetries are often enhanced in the critical region. This has a natural explanation within the framework of the renormalization group.

There are several other common systems in the Ising universality class. For example, a binary fluid, consisting of two components which are miscible above a certain critical temperature, and phase separate below it, has a phase diagram very similar to that of Figure 1.4, with ρ now representing the relative density of one of the components. Similarly, many structural phase transitions in crystals are Ising-like. However, we should issue the warning that such identifications are usually limited to the *static* equilibrium behaviour. As we shall see in Chapter 10, the critical dynamics of these systems may be quite different from each other.

Exercises

1.1 In a model of a binary alloy, each site of the lattice may be occupied by a particle of type A or one of type B. The interactions between the different types of particle are given by $J_{AA}(r - r')$, $J_{BB}(r - r')$ and $J_{AB}(r - r')$. Map this problem to an Ising model by assigning a variable $s(r)$ to each site, which takes the values $+1$ or -1 according to the two cases described above. For what ranges of the interaction parameters should we expect to see critical behaviour of the Ising type?

1.2 An Ising antiferromagnet may be described by the hamiltonian discussed in the text on p.12, with $J(r, r') < 0$. Show that, on a hypercubic lattice, when the dominant interactions are between nearest neighbours, it is possible by a redefinition of variables to map this problem into that of a ferromagnet, so that they are therefore in the same universality class. What happens, for example, on a triangular lattice?

1.3 Real antiferromagnets have magnetic ions with quantum mechanical spins of spin S, not necessarily equal to $\frac{1}{2}$, and exchange interaction J. How would you map (approximately) such a model onto the usual spin-$\frac{1}{2}$ Ising model with some effective exchange interaction J_{eff}? [Hint: compare their behaviours at high temperature.]

1.4 Atoms of a rare gas are adsorbed on a substrate. The adsorbed atoms occupy the sites of a square lattice: however, the radius of the adsorbed atoms is such that neighbouring sites of the square lattice cannot be simultaneously occupied. Show that at high densities and low temperature there are two possible degenerate ground states for this system, and hence deduce the form of the phase diagram in analogy with that of the Ising model. How would the phase transition show up in the structure factor as measured, say, by X-ray or low energy electron diffraction?

2
Mean field theory

Before embarking on an exploration of the modern theories of critical behaviour, it is wise to consider first the more traditional approaches, generally lumped together under the heading of *mean field theory*. Despite the fact that such methods generally fail sufficiently close to the critical point, there are nonetheless several good reasons for their study. Mean field theory is relatively simple to apply in most cases, and often gives a qualitatively correct picture of the phase diagram of a given model. In some cases, where fluctuation effects are suppressed for some physical reason, its predictions are also quantitatively accurate. In a sufficiently large number of spatial dimensions, it gives exact values for the various critical exponents. Moreover, it often serves as an important adjunct to the renormalization group, since the latter by itself may give direct information about the existence and location of phase transitions, but not about the nature of the phases which they separate. For this, further input is necessary, and this often may be provided by applying mean field theory in a region of the phase diagram far from the critical region, where it is applicable.

2.1 The mean field free energy

There are as many derivations of the basic equations of mean field theory as there are books written on the subject, all of varying degrees of rigour and completeness. Here we shall adopt a very simple approach which exposes the essential approximations involved. For illustrative purposes we first consider the ferromagnetic Ising model introduced in Section 1.2. The partition function is

$$Z = \text{Tr} e^{\frac{1}{2}\beta \sum_{r,r'} J(r-r')s(r)s(r') + \beta H \sum_r s(r)}, \qquad (2.1)$$

where $s(r) = \pm 1$. $J(r - r')$ is the ferromagnetic exchange coupling, and H represents a uniform applied magnetic field. By convention, each pair of points (r, r') is counted *twice* in the sum, which accounts for the factor of $\frac{1}{2}$ in front of this term. As usual, $\beta = 1/k_B T$. If the exchange coupling were absent, calculating Z would be trivial, since it would factorise into a product of single site terms, each describing a single Ising spin in a magnetic field. In that case, the magnetisation would simply be given by

$$\langle s \rangle = \frac{\sum_{s=\pm 1} s e^{\beta H s}}{\sum_{s=\pm 1} e^{\beta H s}} = \tanh(\beta H). \qquad (2.2)$$

The central idea of mean field theory is to approximate the interacting case (2.1) by a simpler noninteracting partition function. This must be done in such a way as to encapsulate the essential physics of the problem, and it is at this point that physical insight enters the analysis. In the present case this is straightforward. We know that the applied field H will give rise to some non-zero magnetisation $M = \langle s \rangle$ (and that, in the ferromagnetic phase, this will survive even in the limit $H \to 0$.) Therefore, in some way, we must linearise the term in the hamiltonian which is quadratic in the spin degrees of freedom s, but which preserves this idea of a non-zero magnetisation. One way to do this is to write, as an identity,

$$s(r)s(r') = (M + (s(r) - M))(M + (s(r') - M)), \qquad (2.3)$$

and then to expand this to first order in the deviations $\delta s(r) \equiv s(r) - M$:

$$M^2 + M(s(r) - M) + M(s(r') - M) + O((\delta s)^2). \qquad (2.4)$$

The exponent in (2.1) is then approximated by

$$\tfrac{1}{2}\beta \sum_{r,r'} J(r - r')\Big(M(s(r) + s(r')) - M^2\Big) + \beta H \sum_r s(r) =$$

$$- \tfrac{1}{2} N \beta J M^2 + \beta(JM + H) \sum_r s(r), (2.5)$$

where $J = \sum_{r'} J(r - r')$, and N is the total number of sites. The terms we are neglecting in (2.5) involve the *correlations* between the spins on neighbouring sites. This is clearly rather drastic, since spins at sites within a correlation length of each other will in fact be strongly correlated. We may therefore expect that mean field theory can be quantitatively correct only when the correlation length is fairly small. Later, in Section 2.4, we shall give a more precise criterion for its applicability. However, once the hamiltonian is truncated in this manner, it is then straightforward to calculate the partition function in this approximation, with the result

$$Z \approx e^{-\frac{1}{2}N\beta JM^2} \left[2 \cosh \beta (JM + H)\right]^N. \qquad (2.6)$$

The free energy per site $f \equiv -(k_B T/N) \ln Z$ for a given M is therefore, in the mean field approximation,

$$f_{MF}(M) = \tfrac{1}{2}JM^2 - \beta^{-1} \ln \cosh \beta (JM + H), \qquad (2.7)$$

apart from an irrelevant constant. We have not yet specified M. This is determined by requiring that it minimise† the function $f_{MF}(M)$. This makes physical sense: for a given applied field the most probable value of M is that which minimises the free energy. Setting to zero the derivative of (2.7) with respect to M then gives

$$M = \tanh \beta (JM + H). \qquad (2.8)$$

This equation has a simple physical interpretation. The right hand side is just the magnetisation of a single isolated spin in an external field $JM + H$, which has a part due to the actual external field H, and a part due to the *molecular field* JM produced by the magnetisation of the surrounding spins. There are, in general, several different solutions of (2.8). To decide which one should be chosen, it is useful to consider a plot of $f_{MF}(M)$ against M. For $H = 0$ and small M, f_{MF} has a Taylor expansion

$$f_{MF} = \text{const.} + \tfrac{1}{2}J(1 - \beta J)M^2 + O(M^4), \qquad (2.9)$$

where the coefficient of the M^4 term is positive. Depending on the sign of the quadratic term, the shape of the graph is one of those shown in Figure 2.1. When $T > J/k_B$, the only minimum is at $M = 0$. This then corresponds to the paramagnetic phase. For $T < J/k_B$, we have to choose either M_0 or $-M_0$, and the symmetry $s \rightarrow -s$ is spontaneously broken. Which one is chosen depends on how the limit $H \rightarrow 0$ is taken. For $H \neq 0$ (see Figure 2.2) there is only one minimum, and therefore a unique value for M. Since $Z \sim e^{-\beta N f_{\min}}$, if the thermodynamic limit $N \rightarrow \infty$ is then taken

---

† Other more systematic but less physically transparent approaches to mean field theory show that $f_{MF}(M)$ always gives an upper bound on the true free energy, so it makes sense to minimise this so as to obtain the best such bound.

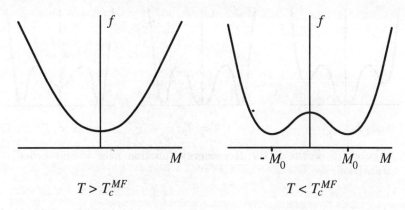

Figure 2.1. Form of the mean field free energy function $f_{MF}(M)$.

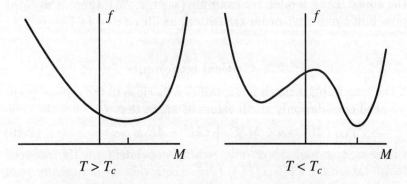

Figure 2.2. Free energy function $f_{MF}(M)$ when $H \neq 0$.

before the limit $H \to 0$, as it should be, either $+M_0$ or $-M_0$ is chosen depending on whether $H \to 0+$ or $H \to 0-$. This shows that we should regard the temperature $T_c^{MF} \equiv J/k_B$ as the critical temperature in the mean field approximation. Note that it depends only on J, which is the sum of all the exchange interactions felt by a given spin, irrespective of their range, anisotropy, and so on. This estimate of the true critical temperature is evidently very crude, and it is almost always an overestimate. This is because fluctuation effects, neglected in mean field theory, tend to have a disordering effect, and therefore suppress the true T_c relative to T_c^{MF}. In sufficiently low dimensions, this suppression can lead to total loss of order at any non-zero temperature, as discussed in

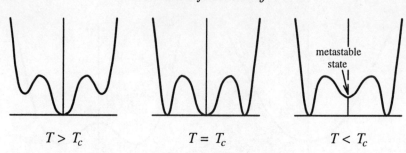

$$T > T_c \qquad\qquad T = T_c \qquad\qquad T < T_c$$

Figure 2.3. Form of the free energy function near a first order transition.

Section 6.1.

If the coefficient of M^4 happens to be negative, as may occur in the spin-1 Ising model, for example (see Ex. 2.5), there is also the possibility of a first-order transition, as illustrated in Figure 2.3.

2.2 Critical exponents

If the transition is continuous, sufficiently close to the critical point we need consider only small values of M, so that f_{MF} has the form

$$f_{MF}(M) = a + btM^2 + cM^4 + dHM + \cdots, \qquad (2.10)$$

where a, b, c and d are only weakly dependent on the reduced temperature $t \equiv (T - T_c^{MF})/T_c^{MF}$. From this we may easily read off the values of the various thermodynamic critical exponents in the mean field approximation. When $H = 0$, and $t < 0$, we have $M_0 \propto (-t)^{1/2}$, so that $\beta = \frac{1}{2}$. If we switch on H, $M \propto cH/at$ as $H \to 0$ (for $t > 0$), so that $\gamma = 1$. Similarly, at $t = 0$, $M \propto (cH/b)^{1/3}$, yielding $\delta = 3$. The analysis of the specific heat is somewhat more subtle. For $t > 0$, the value of f_{MF} at the minimum $M = 0$ is equal to a, while for $t < 0$ it is $a + O\left((bt)^2/c\right)$. There is therefore a *discontinuity* in the specific heat, which is proportional to the second derivative of f_{MF} with respect to t. This is a peculiarity of mean field theory, which does not persist for $d < 4$ when fluctuations are correctly taken into account. The mean field specific heat approaches its discontinuity in a perfectly analytic manner, however, corresponding to a value $\alpha = 0$ for the specific heat exponent.

These results for the mean field exponents show a remarkable

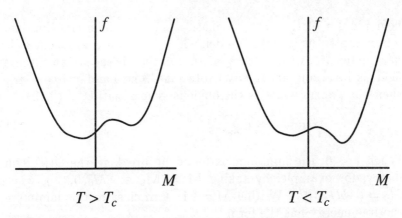

Figure 2.4. Mean field free energy for a system with a cubic invariant.

degree of universality. They are independent not only of the dimensionality, but of almost all the other details of the system. Their actual values depend only on the fact that the mean field free energy may be expanded in the order parameter M in the analytic fashion of (2.10). A little reflection shows that this is in fact the most general analytic form consistent with the symmetries of the problem. In this example, the symmetry is spin reversal $s(r) \rightarrow -s(r)$, which corresponds to $M \rightarrow -M$ and $H \rightarrow -H$. The idea that the effective free energy should be expressible as an analytic function of the order parameter, consistent with the requirements of symmetry of the problem at hand, is the basis of the *Landau theory* of phase transitions. We see that the values of the critical exponents depend only on this assumption, and they are therefore often referred to as the Landau values. While this might seem like a rather trivial observation in the context of the ferromagnetic Ising model discussed above, for systems with more complicated symmetries and order parameters it is very useful to be able to predict the Landau values of the exponents with little or no detailed calculation. We note, in particular, that if the symmetry allows for the presence of invariants which are cubic in the order parameter, such as M^3 in (2.10) above, then Landau theory always predicts a first order transition. This is illustrated in Figure 2.4.

When the system possesses a continuous symmetry, the mean

field picture of the ordered phase is slightly different. Consider, for example, the n-vector model, which has an n-component order parameter M_a, with $1 \leq a \leq n$, and, in the absence of an external field, is invariant under $O(n)$ rotations. The Landau free energy then has a form which is the obvious generalisation of (2.10)

$$f_{MF} = a + bt \sum_a M_a^2 + c(\sum_a M_a^2)^2 + \cdots. \qquad (2.11)$$

When $t < 0$, the minimum is found by breaking the $O(n)$ symmetry, for example, by taking $\mathbf{M} = \mathbf{M_0} \equiv (M_0, 0, \ldots)$, where $M_0 = (-bt/2c)^{1/2}$. Writing $\mathbf{M} = \mathbf{M_0} + \mathbf{m}$ close to this minimum, the free energy has the form

$$f_{MF} = \text{const.} + 2b|t|m_1^2 + \text{higher order terms.} \qquad (2.12)$$

The form of this implies that only the *longitudinal mode* m_1 has a finite correlation length $\propto (2b|t|)^{-1/2}$ for $T < T_c$. The *transverse modes*, corresponding to fluctuations within the minimum energy subspace, have infinite correlation length, and therefore their correlations decay as power laws. These are examples of *Goldstone modes*. *Goldstone's theorem* states that their correlation length remains infinite even when all the fluctuations are taken into account, as long as the symmetry remains broken.

2.3 Mean field theory for the correlation function

The approach described above yields an approximate form for the free energy, and hence for the various thermodynamic quantities and critical exponents which may be derived from this. It is not difficult to modify the method to apply to the correlation functions. For simplicity consider the spin-spin correlation function $G(r)$ in the zero field Ising model in the high temperature phase, when it is equal to

$$\langle s(r)s(0) \rangle \equiv \frac{\text{Tr}\, s(r)s(0)e^{\frac{1}{2}\beta \sum_{r',r''} J(r'-r'')s(r')s(r'')}}{\text{Tr}\, e^{\frac{1}{2}\beta \sum_{r',r''} J(r'-r'')s(r')s(r'')}}. \qquad (2.13)$$

If we insert the identity $1 = \delta_{s(0),1} + \delta_{s(0),-1}$ under the traces in the numerator and denominator, and use the symmetry of the hamiltonian under changing the sign of all the spins, (2.13) may be written as

$$\langle s(r)s(0)\rangle = \frac{\mathrm{Tr}'\, s(r)e^{\frac{1}{2}\beta\sum_{r',r''}J(r'-r'')s(r')s(r'')}}{\mathrm{Tr}'\, e^{\frac{1}{2}\beta\sum_{r',r''}J(r'-r'')s(r')s(r'')}}, \qquad (2.14)$$

where Tr' means the trace over all the other spins, keeping $s(0) = 1$. Thus $G(r)$ is equal to the magnetisation at the site r, given that the spin at the origin is fixed to be $+1$. Interpreted this way, the correlation function is a *response function* to an applied field localised at the origin.

It is now straightforward to go ahead and apply the same mean field approach to calculating this conditional magnetisation as was described in Section 2.1. The only difference is that the problem no longer has translational symmetry, so the mean field magnetisation M depends on r. The self-consistent equation which results from minimising the mean field free energy then becomes

$$M(r) = \tanh\beta\Big(\sum_{r'} J(r-r')M(r')\Big). \qquad (2.15)$$

This is too difficult to solve analytically (although numerically it may be studied quite easily), but, if we are interested only in the large r behaviour, when we know (in the high temperature phase) that $M(r)$ is small, we may expand the right hand side of (2.15) to obtain

$$M(r) = \beta\sum_{r'} J(r-r')M(r')+\text{corrections when } r \text{ is small.} \quad (2.16)$$

The corrections will be quite complicated, but localised around the origin. Insofar as the linearised equation (2.16) is concerned, they act as a localised source for the field $M(r)$, which, for large r, may simply be approximated by a delta function $\propto \delta(r)$. Since the interaction J depends only on the difference $r - r'$, the equation may now be easily solved by Fourier transform:

$$\widetilde{M}(k) = \beta\tilde{J}(k)\widetilde{M}(k) + \text{const.}, \qquad (2.17)$$

where the tilde denotes a Fourier transform, with wave number k. For small k (corresponding to large r), we may expand $\tilde{J}(k) = J(1-R^2k^2)+O(k^4)$. The quantity $R^2 = \sum_r r^2 J(r)/\sum_r J(r)$ gives a measure of the (squared) *range* of the exchange interaction. Solving for $\widetilde{M}(k)$ we then find

$$\widetilde{M}(k) \approx \frac{\text{const.}}{1 - \beta J(1 - R^2k^2)}. \qquad (2.18)$$

Recalling that, within mean field theory, $J = k_B T_c$, the result for the correlation function is thus

$$\tilde{G}(k) \sim \frac{\text{const.}R^{-2}}{k^2 + \xi^{-2}}, \tag{2.19}$$

where $\xi = R t^{-1/2}$. The dependence on R may be checked by considering $\tilde{G}(k = 0)$ which is the susceptibility, given in the mean field approximation of Section 2.2 by $\chi \sim t^{-1}$, independent of R. The inverse Fourier transform of (2.19) now gives the asymptotic behaviour of $G(r) \propto e^{-r/\xi}/r^{(d-1)/2}$, so that we may identify ξ as the correlation length, within the mean field approximation. The result (2.19) of this approximation is called the *Ornstein–Zernicke form* for the correlation function. From the fact that $\xi \propto t^{-1/2}$ we may read off the mean field value of the exponent $\nu = \frac{1}{2}$. At the critical point $G(r)$ is given by the Fourier transform of $1/k^2$, which behaves, for large r, like r^{-d+2}. This yields the mean field value $\eta = 0$.†

2.4 Corrections to mean field theory

The sole approximation made in deriving the mean field results of Section 2.1 was to neglect the terms $\sum_{r'} J(r - r') \delta s(r) \delta s(r')$ in the hamiltonian, where $\delta s = s - M$. This will be self-consistent if the expectation value of this quantity, computed within mean field theory, is small compared with the mean field energy density. From (2.9) we see that the important part of the free energy varies as $f_{MF} \sim J(1 - \beta J)^2$ close to T_c^{MF}, and so the mean field energy density $\partial(\beta f_{MF})/\partial\beta \sim Jt$. Since J is much more short-ranged than the correlation function, we may approximate

$$\sum_{r'} J(r - r')\langle \delta s(r) \delta s(r') \rangle \approx JG(0)$$

$$\approx \frac{\text{const.}J}{R^2} \int_{\text{BZ}} \frac{d^d k}{k^2 + \xi^{-2}}. \tag{2.20}$$

The last integral is, in principle, over the whole of the first Brillouin zone, although we have approximated the integrand by its form valid for small k. This is because we are actually interested

---

† Ornstein–Zernicke theory is sometimes quoted as giving the result $G(r) \sim e^{-r/\xi}/r^{d-2+\eta}$ with $\eta = 0$. This is incorrect: the result quoted in the main text applies for $r \gg \xi$, while the exponent η appears only when $r \ll \xi$.

in extracting the singular behaviour as $\xi \to \infty$, which comes from the neighbourhood of $k = 0$. Because of this, we may further approximate the Brillouin zone by a spherical wave number cut-off $|k| < a^{-1}$, where a is the lattice spacing. We may then write

$$\int_{|k|<a^{-1}} \frac{d^d k}{k^2 + \xi^{-2}} = \int_{|k|<a^{-1}} \frac{d^d k}{k^2} - \xi^{-2} \int_{|k|<a^{-1}} \frac{d^d k}{k^2(k^2 + \xi^{-2})}.$$
(2.21)

The first integral is independent of T and just adds a constant contribution to the energy density from the fluctuations. The second term is convergent as $a \to 0$ if $d < 4$, and then, by dimensional analysis, we see that it is proportional to ξ^{2-d} as $\xi \to \infty$. The temperature-dependent fluctuation contribution to the energy density will therefore be less important than the mean field contribution if $J\xi^{2-d}/R^2 \ll Jt$. Recalling that, in mean field theory, $\xi = Rt^{-1/2}$, this is equivalent to

$$\xi^{4-d} \ll R^4.$$
(2.22)

This (apart from various factors of 2π) is the *Ginzburg criterion*. We see that, if $d < 4$ (which is usually the case experimentally!), as the critical point is approached the correlation length will always grow to a point where (2.22) is violated, and mean field theory therefore fails. However, the point at which it fails depends on the range of the interaction R (as measured in terms of the microscopic distance scale a.) For most systems, for example antiferromagnets, R is of the same order as a, and we therefore expect mean field theory to be invalid by the time the correlation length is a few times a. However, for some systems, e.g. type I superconductors, R is large, of the order of the size of a Cooper pair. For such systems, we expect mean field theory to be a very good approximation, even very close to T_c.

If, on the other hand, $d > 4$, the correction terms discussed above are less singular, and merely change the amplitudes predicted by mean field theory, without modifying the critical exponents. However, it is important to bear in mind that the fluctuations neglected in mean field theory may give rise to contributions, like the first term on the right hand side of (2.21), which are nonsingular but nevertheless large. This causes, for example, a shift in the critical temperature away from its mean field value, even for $d > 4$.

$d = 4$ is called the *upper critical dimension* for the short-range Ising model. Different universality classes have, in general, different upper critical dimensions (see Ex. 2.6), which may be greater or less than the physical number of dimensions. In any case, the fact that most problems simplify in a sufficiently large number of dimensions turns out to provide an important theoretical handle on their renormalization group study.

Exercises

2.1 Derive the mean field equations for the Ising ferromagnet using Feynman's inequality
$$\mathrm{Tr}\, e^{-\mathcal{H}} \geq \mathrm{Tr}\, e^{-\mathcal{H}' - \langle \mathcal{H} - \mathcal{H}' \rangle_{\mathcal{H}'}}$$
where $\mathcal{H}$ is the full hamiltonian and $\mathcal{H}' = h \sum_r s(r)$ is a trial hamiltonian, with h chosen so as to maximise the right hand side.

2.2 Show that the mean field equations are exact in the thermodynamic limit $N \to \infty$ for the *infinite-range* Ising model, with hamiltonian $\mathcal{H} = -(J/N) \sum_{r,r'} s(r)s(r')$, where the sum is over *all* pairs of spins (r, r'). [Hint: write the partition function as a Gaussian integral as in the Appendix (A.1), carry out the trace over the spins, and use the method of steepest descent.]

2.3 Work out the mean field theory for the XY model (see p.12). Show that the mean field values for the critical exponents are the same as those for the Ising model.

2.4 Work out the mean field theory for an Ising antiferromagnet (on a hypercubic lattice) in a uniform magnetic field H. [Hint: you need to introduce two mean field order parameters: the staggered magnetisation and the uniform magnetisation. The latter vanishes when $H = 0$.] What is the phase diagram of this model in mean field theory?

2.5 Work out the mean field theory for the spin-1 Ising model (the Blume–Capel model) given by the hamiltonian in Eq. (4.1). [Hint: do not try to approximate the $s(r)^2$ term.] Show that there is a point in the phase diagram where the coefficient of the M^4 term in the expansion of the mean field free energy vanishes. This corresponds to a tricritical point (see Section 4.1). What are the Landau values for the various standard critical exponents at this point?

2.6 Assuming that the mean field calculation of the correlation function goes through as for the ordinary critical point on p.22, what does the Ginzburg criterion give as the upper critical dimension for Ising tricritical behaviour?

2.7 Repeat the mean field calculation of the correlation function on p.22, for the case $T < T_c$, still in zero applied field. In the relation $\xi \sim \xi_0 |t|^{-\nu}$, what is the ratio of the amplitudes ξ_0 above and below T_c, at least in mean field theory?

2.8 In some liquid crystals, the order parameter is the local electric quadrupole moment, so is represented by a symmetric traceless tensor Q_{ij}. What are the various lower order terms that are allowed, by rotational symmetry, to appear in the Landau free energy? Show that this suggests that the transition into the ordered phase should be first order for this system.

3
The renormalization group idea

In this chapter the basic concepts of the modern approach to equilibrium critical behaviour, conventionally grouped under the title 'renormalization group', are introduced. This terminology is rather unfortunate. The mathematical structure of the procedure, in the sense that it may be said to have any rigorous underpinnings, is certainly not that of a group. Neither is renormalization in quantum field theory an essential element, although it has an intimate connection with some formulations of the renormalization group. In fact, the renormalization group framework may be applied to problems quite unrelated to field theory. The origins of the name may be traced to the particle physics of the 1960s, when it was optimistically hoped that everything in fundamental physics might be explained in terms of symmetries and group theory, rather than dynamics. One of the earliest applications of renormalization group ideas, in fact, was to the rather esoteric subject of the high energy behaviour of renormalized quantum electrodynamics. It took the vision of K. Wilson to realise that these methods had a far wider field of application in the scaling theory of critical phenomena that was being formulated by Fisher, Kadanoff and others in the latter part of the decade. By then, however, the name had become firmly attached to the subject.

Not only are the words 'renormalization' and 'group' examples of unfortunate terminology, the use of the definite article 'the' which usually precedes them is even more confusing. It creates the misleading impression that the renormalization group is a kind of universal machine through which any problem may be processed, producing neat tables of critical exponents at the other end. This is quite false. It cannot be stressed too strongly that the renormalization group is merely a framework, a set of ideas, which has to be adapted to the nature of the problem at hand. In particular,

28

whether or not a renormalization group approach is quantitatively successful depends to a large extent on the nature of the problem, but lack of such success does not necessarily invalidate the qualitative picture it provides.

All renormalization group studies have in common the idea of re-expressing the parameters which define a problem in terms of some other, perhaps simpler, set, while keeping unchanged those physical aspects of the problem which are of interest. This may happen through some kind of coarse-graining of the short-distance degrees of freedom, as in the problem of critical phenomena, where the long-distance physics is of interest. It may represent some kind of modification of the effects of large-scale disturbances, as in fluid turbulence, where the emphasis is on how such fluctuations are fed down to smaller distance scales. Or, in time-dependent problems, such as the dynamics of phase ordering following a quench from a disordered phase, it may correspond to the temporal evolution of the parameters specifying the early-time history, in such a way that the late time properties are left unaltered.

Whatever the motivation, these methods all end up with mathematical equations describing *renormalization group flows* in some complicated parameter space. It is the study of these flows, and what they tell us about the physical problem, which is the essence of renormalization group theory. In the context of equilibrium critical behaviour, this general aspect of the renormalization group appears most directly in the method of *real space* renormalization as applied to lattice spin systems, and it is with this example of the renormalization group in action that we shall therefore begin. It turns out that these real space methods are difficult to control in a quantitative fashion, as there is really no small parameter in which to expand. However, this feature does not weaken the remarkably powerful implications for scaling and universality which arise as quite general properties, and it is these consequences we wish to stress. Later, in Chapter 5, we shall describe a slightly different type of renormalization group, in which a small parameter, related to the number of dimensions of space, does appear, and which may therefore be used to yield systematically improvable quantitative results.

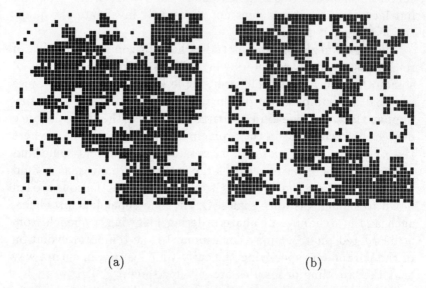

(a)   (b)

Figure 3.1. Typical configuration of the Ising model at $T = T_c$ (a), and (b) the result of a single block spin transformation.

3.1 Block spin transformations

Let us study the accompanying snapshots (Figures 3.1, 3.2) of typical states of the two-dimensional Ising model in zero magnetic field. They were produced by computer simulation. The first picture was taken close to the critical point, and we see clusters of down spins $(s = -1)$ of all sizes. This is what makes the analysis difficult.

Now suppose we put the picture slightly out of focus, so that we can no longer see very well the microscopic details.† A mathematical way to implement this defocussing, or *coarse-graining*, is to make a *block spin transformation*. For definiteness, group the squares into 3×3 blocks, each containing 9 spins. To each block assign a new variable $s' = \pm 1$ whose value will indicate whether the spins in the block are predominantly up or down. The simplest

---

† In a lecture theatre this is rather easy to simulate with an overhead projector. Hyperopic readers should also have no difficulty in performing the experiment.

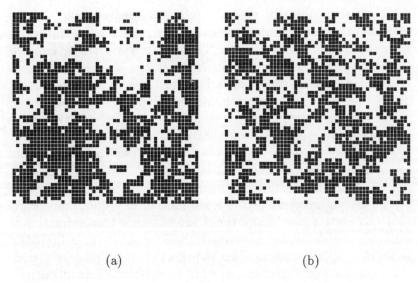

(a)   (b)

Figure 3.2. Same as Figure 3.1, slightly above the critical temperature.

method is to take the 'majority rule' whereby $s' = +1$ if there are
more spins up than down, and *vice versa*. When this is done, and
the whole picture is rescaled by a linear factor of 3 so that the
blocks are the same size as the original squares, we get the second
picture. (It is the same overall size as the first picture because
not all the system is shown there. In fact Figure 3.1a corresponds
to the top left hand corner of Figure 3.1b.) The first thing to be
noticed is that the second picture looks very much like the first.
In fact, they are *statistically* the same, in that Figure 3.1b is an
equally probable configuration for the critical Ising model as is
Figure 3.1a. If we continue this blocking procedure, and start exactly at the critical point with a big enough system, all the pictures
look pretty much the same. This observation illustrates the *scale
invariance* of the critical system. On the other hand, if we start
slightly above the critical temperature (Figure 3.2), although the
original system may look very similar to that in Figure 3.1a, after
a few transformations it soon looks very different.

All this is very qualitative, but it is nevertheless the essential basis of the renormalization group approach. If we were to start from a large sample of typical configurations, we could, by averaging over this sample, calculate all the correlation functions which characterise the system. These are, of course, determined in principle by the hamiltonian $\mathcal{H}(s)$. In the blocked system we could also measure the correlation functions of the block spins s'. In their turn, these correlation functions may be thought of as determined by some new hamiltonian $\mathcal{H}'(s')$. We can always cook up some arbitrarily complicated $\mathcal{H}'$ which will do this, but of course there is no reason to assume that it will be given simply as a sum over nearest neighbour exchange interactions. In general, it will include interactions between arbitrarily distant block spins s'. A basic assumption of the renormalization group, however, is that, no matter how many times the blocking transformation is iterated, the *dominant* interactions will be short-ranged. (Later, in Section 4.3, we shall consider the effect of longer ranged interactions and see how this statement can be made more quantitative.) Rather than attempting to prove the validity of this central assumption for each system of interest, its correctness is best borne out by its consequences of scaling and universality, which have been tested in many real experiments and numerical simulations, as well as being verified for those models which are exactly solvable.

Let us define the block hamiltonian $\mathcal{H}'$ more explicitly. The original system is described by a partition function

$$Z = \mathrm{Tr}_s e^{-\beta \mathcal{H}(s)}. \qquad (3.1)$$

In what follows, we shall always absorb the factor of $\beta = 1/k_B T$ into the definitions of the various parameters in $\mathcal{H}$, such as the exchange coupling J and the magnetic field H. This defines what is called the *reduced* hamiltonian. Our majority rule may be implemented by inserting a projection operator under the trace, as follows. Define, for each block,

$$T(s'; s_1, \ldots, s_9) = \begin{cases} 1, & \text{if } s' \sum_i s_i > 0; \\ 0 & \text{otherwise.} \end{cases} \qquad (3.2)$$

The new hamiltonian is then defined by

$$e^{-\mathcal{H}'(s')} \equiv \mathrm{Tr}_s \prod_{\text{blocks}} T(s'; s_i) \, e^{-\mathcal{H}(s)}. \qquad (3.3)$$

Note that, in particular, because $\sum_{s'} T(s'; s_i) = 1$,

$$\text{Tr}_{s'} e^{-\mathcal{H}'(s')} = \text{Tr}_s e^{-\mathcal{H}(s)}, \tag{3.4}$$

that is, the partition functions Z for the original system and the blocked system are the same. But the above transformation preserves far more than this. Equation 3.3 implies that the whole probability distribution of quantities which depend only on spins $s', s'', s''', \ldots$ at higher levels of blocking will be left invariant. These include all the long wavelength degrees of freedom. Thus the whole of the large distance physics of the problem is left untouched by the renormalization group procedure. The only difference is that it should be expressed in terms of blocked, or renormalized, spins, rather than the original, or bare, spins.

It is useful to think of the couplings in the reduced hamiltonian $\mathcal{H}$ as forming a vector $\{K\} \equiv (K_1, K_2, \ldots)$. In the original model there might have been only one nearest neighbour coupling, say K_1, with all the other $K_i = 0$. But, as discussed above, the renormalization group will in principle generate all other possibilities. We may therefore picture the renormalization group transformation as acting on the space of all possible couplings $\{K\}$:

$$\{K'\} = \mathcal{R}\{K\}. \tag{3.5}$$

In the case of the Ising model, we may divide this very large space into the subspace of even couplings, which multiply interaction terms in the hamiltonian which are invariant under $s(r) \to -s(r)$, and the space of odd couplings, such as an external magnetic field.† If no odd couplings are present in the original model, none should be generated under renormalization.

All of this is rather general. Unfortunately, the sums involved in actually carrying out the trace over the s in (3.3) are intractable, and we must rely on some approximation scheme to proceed further with these block spin methods. It is not the purpose of this book to discuss such schemes, since the approximations involved are difficult to control. Nevertheless, the *qualitative* picture of the renormalization group which we extract from these considerations applies independently of any approximation. It is useful, all the

---

† In the case of a more general symmetry, each subspace corresponds to those interactions which transform according to an irreducible representation of the appropriate symmetry.

Figure 3.3. Blocking transformation for the one-dimensional Ising model.

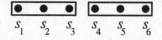

Figure 3.4. Two neighbouring blocks in the one-dimensional Ising model.

same, to examine at least one case in which they may be carried through exactly.

3.2 One-dimensional Ising model

In one dimension, the block spin renormalization group described above may often be carried through explicitly. Consider, for example, the simple zero-field Ising spin chain, with a reduced hamiltonian $\mathcal{H} = -K \sum_i s_i s_{i+1}$. This model may, of course, be solved by other means, for example, by using a transfer matrix, or, even more simply, by defining new variables $\sigma_i \equiv s_i s_{i+1}$. However, for the purposes of illustration, let us consider grouping the sites into blocks, each containing 3 spins, as shown in Figure 3.3. We could use the same majority rule as before, but it is analytically even simpler if we do something rather undemocratic, and count the vote of only the central spin in each block. This corresponds to taking $T(s'; s_1, s_2, s_3) = \delta_{s', s_2}$. The reason that this is justified is that, at very low temperatures, where all the action takes place, all the spins in a given block tend to vote the same way anyway. Thus this renormalization group transformation corresponds simply to performing the trace over the spins at the ends of each block, and leaving the central ones untouched. Such a procedure is called *decimation*. It works very well in one dimension.† Consider two neighbouring blocks, shown in Figure 3.4. Suppose we sum over the spins s_3 and s_4, keeping $s'_1 \equiv s_2$ and $s'_2 \equiv s_5$ fixed. The

† Since in this method the block spins are a subset of the original spins, it leads to the paradoxical result that the spin correlation function is constant at the critical point, which is incorrect in higher dimensions.

factors in the partition sum involving these degrees of freedom are

$$e^{Ks_1' s_3} \; e^{Ks_3 \, s_4} \; e^{Ks_4 \, s_2'}. \tag{3.6}$$

Write $e^{Ks_3 s_4} = \cosh K(1+x s_3 s_4)$, where $x \equiv \tanh K$, and similarly for the other two factors, giving

$$(\cosh K)^3 (1 + x s_1' s_3)(1 + x s_3 s_4)(1 + x s_4 s_2'). \tag{3.7}$$

Now imagine expanding out this expression. When we perform the sum over $s_3, s_4 = \pm 1$, only terms with an even power of these variables will survive. We are therefore left with only

$$2^2 (\cosh K)^3 (1 + x^3 s_1' s_2'), \tag{3.8}$$

which, apart from the constant outside, has the form of a nearest neighbour Boltzmann factor $e^{K' s_1' s_2'}$, with

$$K' = \tanh^{-1} \left[(\tanh K)^3 \right]. \tag{3.9}$$

The partition function of the whole system may thus be written in the required form $Z = \mathrm{Tr}_{s'} e^{-\mathcal{H}'(s')}$, where

$$\mathcal{H}'(s') = Ng(K) - K' \sum_i s_i' s_{i+1}', \tag{3.10}$$

where N is the total number of original sites, and

$$g(K) = -\tfrac{1}{3} \ln \left[\frac{(\cosh K)^3}{\cosh K'} \right] - \tfrac{2}{3} \ln 2. \tag{3.11}$$

The renormalized hamiltonian, has, in this case, the same form as the original one, with a renormalized value of the coupling K', apart from a term independent of the s_i'. This additional term proportional to $g(K)$ does not affect the calculation of any expectation values, but it will enter into a calculation of the total free energy. It represents the contribution to the free energy from the short wavelength degrees of freedom which have been traced out. As we shall see in Sections 3.4 and 3.9, it plays an important role in the further development of the theory.

The content of the renormalization group transformation is expressed by the *renormalization group equation* (3.9). This is much more easily analysed in terms of the variable $x = \tanh K$, for which the renormalization group equation is simply $x' = x^3$. Recalling that K contains a factor $1/k_B T$, we see that high temperatures correspond to $x \to 0+$, and low temperatures to $x \to 1-$. Now suppose we iterate the process. Unless we begin with x exactly

$$K = \text{infinity} \qquad\qquad\qquad\qquad\qquad K = 0$$
$$T = 0 \qquad\qquad\qquad\qquad\qquad\qquad T = \text{infinity}$$

Figure 3.5. RG flow for the one-dimensional Ising model.

equal to 1 ($T = 0$), it ultimately approaches zero. This means that the long distance degrees of freedom are described by a hamiltonian where the effective temperature is high, and we expect such a system to be in a paramagnetic state, with a finite correlation length. Since any system with $x < 1$ ultimately renormalizes into this region, we conclude that this whole region is paramagnetic. Only exactly at zero temperature is this not true. In terms of the renormalization group equation, we may say that there are two *fixed points*: the one at $T = 0$ is unstable, since any perturbation away from this is amplified by the renormalization group; the fixed point at $T = \infty$ ($x = 0$) is stable, and is the attractive fixed point for the whole region $0 \le x < 1$. Every point in this region is therefore in the same phase, which is paramagnetic.† The *renormalization group flows* go from the unstable fixed point to the stable one, as shown in Figure 3.5.

This reflects a well-known fact about the one-dimensional systems with short-range interactions: they cannot be in an ordered state for $T > 0$, and therefore can have no true phase transition. This tendency towards the disappearance of order as the number of dimensions is lowered is characteristic of all systems. The dimension d_l such that systems in a given universality class have no phase transition for $d \le d_l$ is called the *lower critical dimension*. For systems like the Ising model with discrete symmetries, $d_l = 1$. As we shall see in Section 6.1, for continuous symmetries $d_l = 2$ in general. This behaviour makes the above example rather uninteresting for our purposes, but, on the other hand, the analysis was rather simple. To illustrate how the renormalization group can be useful even when there is no true critical behaviour, however, let us use the above results to compute the correlation length ξ.

---

† Notice that we needed to supply additional physical input to describe the *nature* of the phase, which is not determined solely by the renormalization group.

The correlation length has, of course, the dimensions of length, but to express it as a pure number we may measure it in units of the lattice spacing a. In these units, it may depend on only the reduced coupling K. After performing the renormalization group transformation, the long distance physics is preserved, and so the dimensionful correlation length must remain the same. However, the lattice spacing has increased by a factor of $b = 3$ (in this particular example). Thus the dimensionless correlation length transforms according to

$$\xi(x') = b^{-1}\xi(x), \tag{3.12}$$

where $x' = x^b$. This has the solution

$$\xi(x) = \frac{\text{const.}}{\ln x} = \frac{\text{const.}}{\ln \tanh K}, \tag{3.13}$$

which is the exact result for the one-dimensional Ising model. We see that ξ is finite, as expected, but that, as $T \to 0$, $\xi \propto e^{\text{const.}/T}$, so that it grows very large as the system approaches perfect ordering at $T = 0$.

Higher dimensions

The kind of analytic block spin renormalization group we have discussed above is no longer feasible for $d > 1$. Progress is possible by making various simplifying approximations, but, even then, the calculations rapidly become cumbersome and give little insight into the physics. Rather than pursue this subject in detail, then, we shall be content with making some general observations.

In one dimension it follows from (3.9) that at low temperature ($K \to \infty$) the renormalization group equation simplifies to $K' \sim K - \text{const}$. This may be easily understood on physical grounds: at low temperature the spins in the blocks are almost always aligned in the same state. The interaction between adjacent blocks is mediated by their boundary spins (s_3 and s_4 in Figure 3.4). Thus we may write

$$K' \sim K \langle s_3 \rangle_{s_1'=1} \langle s_4 \rangle_{s_2'=1}, \tag{3.14}$$

where $\langle s_3 \rangle_{s_1'=1}$ is the magnetisation of the boundary spin, given that the block spin is $+1$. At low temperature, this is unity, so that $K' \sim K$.

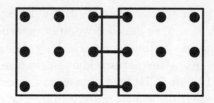

Figure 3.6. Neighbouring blocks in two dimensions.

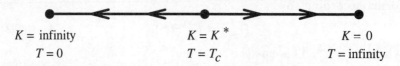

$K = \text{infinity}$ $K = K^*$ $K = 0$

$T = 0$ $T = T_c$ $T = \text{infinity}$

Figure 3.7. Schematic RG flow in the Ising model for $d > 1$.

Now consider the case of two neighbouring blocks in two dimensions, as illustrated in Figure 3.6. The interaction between neighbouring blocks is now mediated by three nearest neighbour bonds, so that, as $K \to \infty$, we expect that $K' \sim 3K$. In the case of d dimensions, with a length rescaling factor b, this generalises to

$$K' \sim b^{d-1} K \qquad \text{as} \qquad K \to \infty. \qquad (3.15)$$

This has the important consequence that, for $d > 1$, $K' > K$, so that the zero-temperature fixed point at $K^{-1} = 0$ is locally stable. On the other hand, at high temperatures the system has to be in a paramagnetic phase, so the high-temperature fixed point must also be stable. Therefore, to the extent that we can think of the renormalization group flows as being unidimensional, there must exist an unstable fixed point $K = K^*$ in between them, as shown in Figure 3.7. This fixed point corresponds to the critical point. To see this, imagine calculating the correlation length using $\xi(K) = b\xi(K')$. Suppose that at some fixed high temperature, corresponding to a reduced coupling K_0, $\xi = \xi_0 = O(1)$. Starting at some $K < K^*$, it will take a certain number $n(K)$ of iterations of the renormalization group before we reach the vicinity of K_0. Then $\xi(K) = \xi_0 b^{n(K)}$. As K approaches K^*, the amount by which K changes under each iteration is initially very small, so that $n(K)$ becomes large. As $K \to K^*$, $n(K) \to \infty$, so that $\xi(K) \to \infty$,

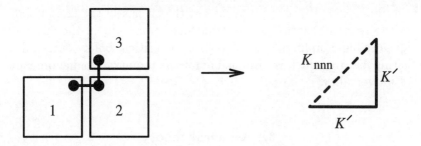

Figure 3.8. Generation of next-nearest neighbour coupling.

indicating a critical point.

In fact, knowing how K' depends on K close to the fixed point, we may calculate the critical exponent ν. Suppose that $K' = \mathcal{R}(K)$, where $K^* = \mathcal{R}(K^*)$. When $K - K^*$ is small, we may then write

$$K' \approx \mathcal{R}(K^*) + (K - K^*)\mathcal{R}'(K^*) = K^* + b^y(K - K^*), \quad (3.16)$$

which defines the quantity $y \equiv \ln \mathcal{R}'(K^*)/\ln b$. Now, close to the critical point, we expect from p.7 that $\xi(K) \sim A(K - K^*)^{-\nu}$. Using $\xi(K) = b\xi(K')$, it then follows that

$$A(K - K^*)^{-\nu} = bA(K' - K^*)^{-\nu} = bA\left[b^y(K - K^*)\right]^{-\nu}, \quad (3.17)$$

which is only possible if $\nu = 1/y$. This is an example of a general result (see Section 3.5) that critical exponents are given in terms of the *derivatives* of the renormalization group transformation at the fixed point. Indeed, we see from the above that we may trace the very existence of a critical exponent to the assumption of differentiability of the renormalization group transformation at the fixed point.

However, our analysis has, up to now, been lacking in one important respect, which does not arise in one dimension. Consider the effect of summing over the corner spin of block 2 in Figure 3.8: This couples to spins in blocks 1 and 3, so that it will generate an effective coupling between the respective block spins $1'$ and $3'$. On the blocked lattice, this will be a next-nearest neighbour coupling. In fact, as discussed earlier, all possible further neighbour couplings will now appear, and the one-dimensional picture of the renormalization group flows in Figure 3.7 is therefore a gross over-simplification. However, it turns out that it is still the fixed points of the renormalization group transformation, now in the space of all possible couplings, which control the critical behaviour, and it is this feature which is responsible for the remarkable phenomenon of universality.

3.3 General theory

In this section we shall examine the consequences of the very general assumption that there exists a renormalization group fixed point in the space of all possible couplings. The transformation has the form $\{K'\} = \mathcal{R}(\{K\})$, where $\mathcal{R}$ will depend, in general, on the specific transformation chosen, and, in particular, on the length rescaling parameter b. Suppose there is a fixed point at $\{K\} = \{K^*\}$. As in the single variable case, we shall assume that $\mathcal{R}$ is differentiable at the fixed point, so that the renormalization group equations, linearised about the fixed point, are

$$K'_a - K^*_a \sim \sum_b T_{ab}(K_b - K^*_b), \tag{3.18}$$

where $T_{ab} = \partial K'_a / \partial K_b |_{K=K^*}$. Denote the eigenvalues of the matrix $\mathbf{T}$ by λ^i, and its left eigenvectors by $\{e^i\}$, so that

$$\sum_a e^i_a T_{ab} = \lambda^i e^i_b. \tag{3.19}$$

Note that we have no reason to suppose that $\mathbf{T}$ is symmetric, so that its left eigenvectors are not in general the same as the corresponding right eigenvectors. In fact, we are not even entitled to assume that the eigenvalues λ^i are real, but, as may be seen from the subsequent discussion, strange things would happen if they were not.†

Now define *scaling variables*‡ $u_i \equiv \sum_a e^i_a (K_a - K^*_a)$, which are linear combinations of the deviations $K_a - K^*_a$ from the fixed point

---

† In some random systems, the irrelevant eigenvalues, which correspond to corrections to scaling (see Section 3.6) may occur in complex conjugate pairs.

‡ See the discussion at the foot of p.53 on terminology.

which transform *multiplicatively* near the fixed point:

$$u_i' = \sum_a e_a^i (K_a' - K_a^*) = \sum_{a,b} e_a^i T_{ab} (K_b - K_b^*)$$

$$= \sum_b \lambda^i e_b^i (K_b - K_b^*) = \lambda^i u_i. \tag{3.20}$$

It is convenient to define the quantities y_i by $\lambda^i = b^{y_i}$. The y_i are called *renormalization group eigenvalues*, and will turn out to be related to the critical exponents. We may distinguish three cases:

- If $y_i > 0$, u_i is said to be *relevant*: repeated renormalization group iterations drive it away from its fixed point value.
- If $y_i < 0$, u_i is *irrelevant*: if we start sufficiently close to the fixed point, u_i will iterate towards zero.
- If $y_i = 0$, u_i is *marginal*. In this case, we cannot tell from the linearised equations whether u_i will move away from the fixed point or towards it. An example of this interesting case will be discussed later in Section 5.6.

Let us now consider a fixed point which has n relevant eigenvalues. For convenience, we may imagine the space near the fixed point as having n' dimensions in all (although strictly speaking this is infinite). There will then be $(n' - n)$ irrelevant eigenvalues, so, in the vicinity of the fixed point, there will be an $(n' - n)$-dimensional hypersurface of points attracted into the fixed point. Near the fixed point, this is just the linear space spanned by the irrelevant eigenvectors, but, by continuity, we expect this hypersurface to exist in some finite region around the fixed point. It is called the *critical surface*, since the long distance properties of each system corresponding to a point on this surface will be controlled by the fixed point, at which, by the same arguments as in Section 3.2, the correlation length will be infinite. Now the coupling constants K_a will depend in some complicated manner on the physical parameters like the temperature, pressure or magnetic field which the experimentalist may vary. We shall refer to these as 'knobs' which the experimentalist may adjust. In order to end up on the $(n'-n)$-dimensional surface attracted into this fixed point, she must therefore adjust exactly n knobs.

In the example of the ferromagnetic Ising model, two knobs must be adjusted (T and H) to bring the system to its critical

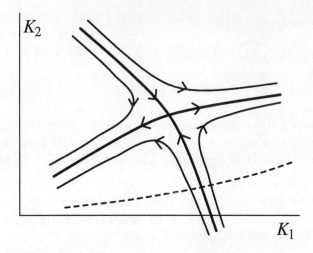

Figure 3.9. RG flows in a two-dimensional example.

point. The same is true of a simple fluid (e.g. pressure and temperature). Thus we expect the fixed point corresponding to this universality class to have just two relevant scaling variables. Since the critical point occurs at zero field in the Ising model, the corresponding fixed point must also occur in the subspace where all the odd couplings vanish. This means that the matrix $\mathbf{T}$ must be block diagonal, having no elements which connect the even and odd subspaces, and we may therefore classify all its eigenvectors as being either even or odd under the symmetry $s \rightarrow -s$. Thus one of the relevant variables must be temperature-like, and lie in the even subspace – this is called the *thermal* scaling variable – and the other, the *magnetic* scaling variable, must lie in the odd subspace.

In order to illustrate the significance of the irrelevant fields, let us restrict attention to flows in the even subspace, and, for simplicity, reduce the number of dimensions of this subspace to two, so that it is parametrised by (K_1, K_2). For example, K_1 might be the reduced nearest neighbour coupling, and K_2 the next-nearest neighbour coupling. There is then just one relevant and one irrelevant eigenvalue. The topology of the flows near the fixed point is shown in Figure 3.9. There is a one-dimensional curve of points attracted into the fixed point. This is the critical surface in this toy example. The flows near the fixed point must have the hyperbolic form shown in the figure by continuity. Thus the critical surface, in this example, acts as a separatrix, dividing the region of points which flow to large values of the K_a (ultimately to zero temperature) from those flowing to small K_a (ultimately to the high-temperature fixed point at $K_a = 0$). In the model with nearest neighbour couplings only, as the temperature is varied we move along the axis $K_2 = 0$. The point where this line meets the critical surface then defines the critical reduced coupling K_{1c}, corresponding to the critical temperature of the nearest neighbour model, since points with $K < K_{1c}$ and $K > K_{1c}$ end up at the high- or low-temperature fixed points respectively. At $K = K_{1c}$, the renormalization group trajectories flow into the critical fixed point, which means that the long distance behaviour at the critical point is the same as that of the fixed point.

However, we may equally well consider a model with a next-nearest neighbour coupling in addition. Then, as we change the temperature, we move along some other curve in coupling constant space, indicated by the dashed line in Figure 3.9. The critical point of this model occurs where this line intersects the critical surface. But the large distance behaviour in this case will be similar to that in the case of a simple nearest neighbour coupling, because they are both controlled by the same fixed point.

This argument, suitably generalised to the case of an infinite-dimensional coupling constant space, is the simple explanation of the phenomenon of universality. A universality class consists of all those critical models which flow into a particular fixed point. To each universality class will correspond a different critical fixed point. However, in order to understand precisely which quantities are universal, we need to understand just what information the fixed point actually provides about the critical theory.

3.4 Scaling behaviour of the free energy

For definiteness, let us continue to consider the universality class of the critical short-range Ising model. As discussed in the previous section, there is a relevant thermal scaling variable u_t, with eigenvalue y_t, and a relevant magnetic scaling variable u_h, with eigenvalue y_h. In addition, there will be an infinite number of irrelevant variables $u_3, \ldots$ The critical point of the model in which we are interested will, in general, lie some finite distance away from the fixed point in coupling constant space. However, only a *finite* number of renormalization group iterations will be required to bring the renormalized theory to the vicinity of the fixed point, where the linearised version of the renormalization group equations is valid. The values u_i of the scaling fields at this point will therefore depend *analytically* on the deviations (t, h) of the original theory from its critical point. This is because the renormalization group transformation itself is analytic, and so therefore is also the result of a finite number of iterations. The relevant variables (u_t, u_h) must also vanish when $t = h = 0$, so that, by symmetry, they must have the form

$$u_t = t/t_0 + O(t^2, h^2) \tag{3.21}$$

$$u_h = h/h_0 + O(th), \tag{3.22}$$

where t_0 and h_0 are non-universal constants. Therefore, close to the critical point, we may take u_t and u_h to be proportional to t and h respectively.†

Recall that one of the properties of the renormalization group transformation is that it preserves the partition function:

$$Z = \mathrm{Tr}_s \, e^{-\mathcal{H}(s)} = \mathrm{Tr}_{s'} \, e^{-\mathcal{H}'(s')}. \tag{3.23}$$

Consider the *reduced free energy per site*, $f(\{K\}) \equiv -N^{-1} \ln Z$, as a function of the couplings $\{K\}$. Under renormalization, the couplings flow according to the renormalization group equations, but in addition a constant term $Ng(\{K\})$ is added to the free energy, as in (3.10). Thus

$$e^{-Nf(\{K\})} = e^{-Ng(\{K\}) - N'f(\{K'\})}, \tag{3.24}$$

where $N' = b^{-d}N$ is the total number of blocks. This gives the fundamental transformation law for the free energy per site:

$$\boxed{f(\{K\}) = g(\{K\}) + b^{-d}f(\{K'\})} \tag{3.25}$$

Notice that the free energy, unlike the correlation length, transforms inhomogeneously under the renormalization group. However, if we are interested in extracting only the *singular* behaviour

---

† Note that for fixed points with more relevant variables, e.g. at a tricritical point (Section 4.1), there is no reason to suppose that the scaling fields are simply proportional to the experimentalist's 'knobs' unless there is some symmetry enforcing this.

of f, for the purpose, for example, of calculating the critical exponents, we may in fact drop the inhomogeneous g term. Physically this is because it originates from summing over the short wavelength degrees of freedom within each block, so that $g(\{K\})$ should be an analytic function of the K_a, even at the critical point. In this manner we obtain a *homogeneous* transformation law for the *singular part* of the free energy f_s

$$f_s(\{K\}) = b^{-d} f_s(\{K'\}).\qquad(3.26)$$

Close to the fixed point, we may write this in terms of the scaling variables

$$f_s(u_t, u_h) = b^{-d} f_s(b^{y_t} u_t, b^{y_h} u_h) = b^{-nd} f_s(b^{ny_t} u_t, b^{ny_h} u_h),\quad(3.27)$$

where, for the time being, the irrelevant variables $u_3, \ldots$ are ignored. In the last expression, we have iterated the renormalization group n times. Since the variables u_t and u_h are growing under this iteration, we cannot make n too large, or the linear approximation to the renormalization group equations would eventually break down. So let us choose to halt the iteration at the point where $|b^{ny_t} u_t| = u_{t0}$, where u_{t0} is arbitrary but fixed, and sufficiently small so that the linear approximation is still valid. Solving this equation for n, we then find, after a little algebra, that

$$f_s(u_t, u_h) = |u_t/u_{t0}|^{d/y_t} f_s\left(\pm u_{t0}, u_h |u_t/u_{t0}|^{-y_h/y_t}\right).\qquad(3.28)$$

Rewriting this in terms of the reduced physical variables t and h, we see that u_{t0} may be incorporated into a redefinition of the scale factor t_0, and that

$$f_s(t, h) = |t/t_0|^{d/y_t} \, \Phi\left(\frac{h/h_0}{|t/t_0|^{y_h/y_t}}\right),\qquad(3.29)$$

where Φ is a *scaling function*.† This function might appear to depend on u_{t0}, but since the left hand side of (3.29) cannot, this is illusory, and, in fact, such scaling functions turn out to be *universal*. The only dependence on the particular system is through the *scale factors* t_0 and h_0.

3.5 Critical exponents

From the scaling law (3.29) for the singular part of the free energy follow all the thermodynamic exponents:

† Note that there are, in fact, different scaling functions for $t > 0$ and $t < 0$.

- Specific heat $\partial^2 f / \partial t^2 |_{h=0} \propto |t|^{d/y_t - 2}$, so that

$$\boxed{\alpha = 2 - d/y_t} \tag{3.30}$$

- Spontaneous magnetisation $\partial f / \partial h |_{h=0} \propto (-t)^{(d-y_h)/y_t}$, so that

$$\boxed{\beta = \frac{d - y_h}{y_t}} \tag{3.31}$$

- Susceptibility $\partial^2 f / \partial h^2 |_{h=0} \propto |t|^{(d-2y_h)/y_t}$, so that

$$\boxed{\gamma = \frac{2y_h - d}{y_t}} \tag{3.32}$$

- To get δ we must work a little harder: we have

$$M = \frac{\partial f}{\partial h} = |t/t_0|^{(d-y_h)/y_t} \, \Phi' \left(\frac{h/h_0}{|t/t_0|^{y_h/y_t}} \right). \tag{3.33}$$

When inverted to express h as a function of M, this is called the scaling form of the equation of state (Widom scaling). For M to have a finite limit as $t \to 0$, $\Phi'(x)$ must behave like $x^{d/y_h - 1}$ as $x \to \infty$. Thus, at $t = 0$, $M \propto h^{d/y_h - 1}$, or

$$\boxed{\delta = \frac{y_h}{d - y_h}} \tag{3.34}$$

We see that the four principal thermodynamic exponents are given in terms of the two renormalization group eigenvalues y_t and y_h. This means that there must exist *scaling relations* between them. Examples are

$$\alpha + 2\beta + \gamma = 2, \tag{3.35}$$

$$\alpha + \beta(1 + \delta) = 2. \tag{3.36}$$

These relations, among others, were postulated before the advent of the renormalization group. Many of them may be proved rigorously as inequalities. Although the relations between the critical exponents and the renormalization group eigenvalues were established above for the case of the Ising universality class, similar equations hold for any universality class with a single relevant thermal eigenvalue and a relevant symmetry-breaking field. However, for more complicated cases, describing for example multicritical points (see Sections 4.1 and 4.2), the corresponding relations are more complex and should be derived from first principles as above.

Role of the rescaling factor b

In the previous section, we saw how the various thermodynamic exponents are related to the eigenvalues b^{y_i} of the matrix T_{ab} of derivatives of the renormalization group transformation at the fixed point. Since the length rescaling factor b enters explicitly into this calculation, one might legitimately ask whether it plays a role in the final values for the exponents. The answer is, of course, that it cannot, since the exponents are properties of the system under consideration, rather than the particular renormalization group transformation which is applied. In fact, the renormalization group transformation contains b implicitly, and it must therefore happen that, *if* the transformation is carried out exactly, the physical observables such as the exponents and scaling functions are independent of b. In the few cases where an exact renormalization group solution is available (for example in one dimension) or when it may be cast in the form of a controlled approximation (as in the ϵ-expansion), the independence of the final results of the details of the transformation may indeed be verified. In the case of block spin transformations, however, which usually demand uncontrolled approximations, the *approximate* values obtained for the exponents do exhibit weak dependence on b and other details of the transformation.

Although for a block spin transformation the possible values of the rescaling factor b are strongly limited by the requirement that the blocked lattice should have the same structure as the original, as we shall see later there are other forms of the renormalization group in which b may be arbitrary. In these cases it is often helpful to consider the limit of an *infinitesimal* transformation, when $b = 1 + \delta\ell$, with $\delta\ell \ll 1$. In this case, the couplings will also transform infinitesimally

$$K_a \to K_a + (dK_a/d\ell)\delta\ell + O(\delta\ell^2), \qquad (3.37)$$

and the renormalization group equations take the differential form

$$dK_a/d\ell = -\beta_a(\{K\}), \qquad (3.38)$$

where the functions β_a are called the renormalization group beta functions. (The minus sign in (3.38) is included so as to make contact with the conventional definition of the beta function in quantum field theory, $\beta_a = \kappa \partial K_a/\partial\kappa$, where κ is a wave number, rather than a length scale.) In this infinitesimal form of the renormalization group, the fixed points now correspond to the *zeroes* of the beta functions. The matrix of derivatives at the fixed point is now $T_{ab} = \delta_{ab} + (\partial \beta_a / \partial K_b)\delta\ell$, with eigenvalues $(1+\delta\ell)^{y_i} \sim 1+y_i\delta\ell$. Hence the y_i are simply the eigenvalues of the matrix $-\partial\beta_a/\partial K_b$, evaluated at the zero of the beta functions. These infinitesimal renormalization group transformations will play a central role in Chapter 5.

3.6 Irrelevant eigenvalues

Suppose that we have an irrelevant scaling variable u_3, with eigenvalue $y_3 < 0$. If it is in the even subspace, we may assume that its initial value depends analytically on t and h, and is therefore of the form

$$u_3 = u_3^0 + at + bh^2 + \cdots, \tag{3.39}$$

where $a, b, \ldots$ are constants. Unlike the case of a relevant variable, however, we may not assume that $u_3^0 = 0$. Close to the critical point, we may initially ignore the higher order terms and set $u_3 = u_3^0$, which has some non-universal value. Repeating our calculation in Section 3.4 of the free energy, we now find that

$$f_s(t, h) \sim |t|^{d/y_t} \, \Phi\left(h|t|^{-y_h/y_t}, u_3^0|t|^{|y_3|/y_t}\right). \tag{3.40}$$

Since $u_3^0|t|^{|y_3|/y_t}$ is small as $t \to 0$, and the right hand side represents the free energy evaluated away from the critical point, it seems reasonable to assume that it is an analytic function of its arguments, and that, in particular, we expand it in a Taylor series in $u_3^0|t|^{|y_3|/y_t}$. Taking $h = 0$ for clarity, we then find that

$$f_s = |t|^{d/y_t} \left(A_1 + A_2 u_3^0|t|^{|y_3|/y_t} + \cdots\right), \tag{3.41}$$

where A_1, A_2, *etc.* are non-universal constants. We see that the leading effect of such irrelevant variables is to give rise to *correction to scaling* terms.† Since in a real system the coefficient of these terms may be quite large, it may be difficult to observe the true asymptotic exponents except very close to the critical point. In that case, a fit to the data which does not include such correction

---

† These are sometimes called *confluent singularities*, usually in the context of extracting critical behaviour from high-temperature expansions.

to scaling terms may erroneously lead to the conclusion that the exponents extracted in this way are non-universal. On the other hand, fitting data with the correction to scaling terms included introduces many more parameters, and thus requires very high quality data to obtain a meaningful fit.

As well as the non-analytic correction to scaling terms of the type shown in (3.41), there are also corrections that come from the higher order dependence of the starting values of the u_i on t and h. Typically, they lead to corrections which are down by relative integral powers of t and h on the leading terms, and are thus called analytic corrections. Since the exponent $|y_3|/y_t$ is around 0.5 for many three-dimensional systems, the non-analytic corrections are more important close to the critical point.

Implicit in the above discussion was the assumption that the Taylor expansion of the right hand side of (3.41) exists, and the limit $u_3 \to 0$ is smooth and well-defined. However, there are situations, several of which will arise later (see Sections (4.4, 8.4, 9.4)), when this is not true. In such a case, u_3 is referred to as a *dangerous* irrelevant variable, and we need more information about the dependence on u_3 to infer from (3.41) the true behaviour of the free energy in the critical region.

3.7 Scaling for the correlation functions

The transformation law for the free energy (3.25) relied only on the property of the renormalization group that it preserves the partition function. However, as we have seen in Section 3.1, the renormalization group in fact preserves the whole probability measure of the long wavelength degrees of freedom, and therefore analogous transformation laws should apply to the large distance behaviour of the correlation functions. This is indeed the case. As an example we shall consider the spin-spin 2-point correlation function in the Ising model, defined by

$$G(r_1 - r_2, \mathcal{H}) \equiv \langle s(r_1)s(r_2)\rangle_{\mathcal{H}} - \langle s(r_1)\rangle_{\mathcal{H}}\langle s(r_2)\rangle_{\mathcal{H}}, \qquad (3.42)$$

where we have emphasised the dependence of G on the parameters in the hamiltonian $\mathcal{H}$. We may also obtain G by adding a non-uniform magnetic field to the hamiltonian $\mathcal{H} \to \mathcal{H} - \sum_r h(r)s(r)$

and differentiating the free energy with respect to $h(r)$:

$$G(r_1 - r_2) = \frac{\partial^2}{\partial h(r_1)\partial h(r_2)} \ln Z\{h\}\bigg|_{h(r)=0}. \qquad (3.43)$$

We now suppose that $h(r)$ varies significantly only over distances much large than the size ba of the blocks, and imagine applying the same type of block spin renormalization group as in Section 3.1. If the hamiltonian $\mathcal{H}$ (which is close to the fixed point hamiltonian $\mathcal{H}^*$, since we are interested in the critical region) contains only short-range interactions, then, in performing the block spin transformation in one region, around r_1 say, we may effectively ignore the fact that $h(r)$ is actually slowly varying, and assume that it transforms in the same manner as would a weak *uniform* field $h = h(r_1)$. According to this argument, the renormalized hamiltonian is therefore of the same form

$$\mathcal{H}'(s') - \sum_{r'} h'(r')s'(r'), \qquad (3.44)$$

where $h'(r') = b^{y_h}h(r)$. Since the renormalization group preserves the entire partition function, however,

$$\frac{\partial^2 \ln Z'(h')}{\partial h'(r_1')\partial h'(r_2')} = \frac{\partial^2 \ln Z(h)}{\partial h'(r_1')\partial h'(r_2')}. \qquad (3.45)$$

Let us examine the meaning of each side of this equation. The left hand side is just the correlation function of the block spins in the ensemble defined by the renormalized hamiltonian $\mathcal{H}'$. However, in units of the lattice spacing, the distance between the points has been reduced by a factor of b. Thus the left hand side is simply $G((r_1 - r_2)/b, \mathcal{H}')$. The right hand side is more tricky. Making an infinitesimal local change $h'(r_1') \to h'(r_1') + \delta h'(r_1')$ within block number 1 corresponds to changing *all* the fields $h(r_i)$ acting on the spins within this block, by an amount $\delta h(r_i) = b^{-y_h}\delta h'(r_1')$. Thus the right hand side is

$$b^{-2y_h}\langle(s_1^{(1)} + s_2^{(1)} + \cdots)(s_1^{(2)} + s_2^{(2)} + \cdots)\rangle_{\mathcal{H}}, \qquad (3.46)$$

where the spins in blocks 1 and 2 are labelled by $s_i^{(1)}$ and $s_i^{(2)}$ respectively, and the subscript denotes that this correlation function is to be evaluated with respect to the original hamiltonian $\mathcal{H}$. Since there are b^d spins in each block, (3.46) may be expanded as a sum of b^{2d} two-point correlations. If $|r_1 - r_2|$ is much larger

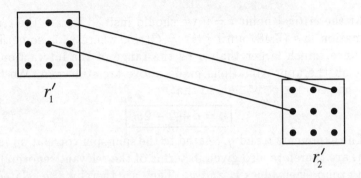

Figure 3.10. Correlations between spins in two distant blocks.

than b, all these correlation functions are, however, numerically almost the same (see Figure 3.10). The transformation law for the correlation function, close to the fixed point, is therefore

$$G((r_1 - r_2)/b, \mathcal{H}') = b^{2(d-y_h)} G(r_1 - r_2, \mathcal{H}). \qquad (3.47)$$

If the interactions are isotropic (respecting, for example, the rotational symmetries of the lattice), then, at large enough distances, the 2-point correlation function in fact depends only on the distance $r = |r_1 - r_2|$ and not on the relative orientation of the two points. This is because scaling fields which break the full rotational symmetry down to that of the lattice are irrelevant.

Setting the uniform magnetic field $h = 0$ for clarity, near the critical point

$$G(r, t) = b^{-2(d-y_h)} G(r/b, b^{y_t} t). \qquad (3.48)$$

We may now iterate this equation n times, as we did for the free energy, stopping at a point where $b^{ny_t}(t/t_0) = 1$. After a little algebra, it follows that the correlation function has the scaling form

$$G(r, t) = |t/t_0|^{2(d-y_h)/y_t} \, \Psi \left(r/|t/t_0|^{-1/y_t} \right). \qquad (3.49)$$

For sufficiently large r, we expect G to decay as $e^{-r/\xi}$, since this also corresponds to large t and the region where the mean field result of Section 2.3 should apply. From (3.49) we then identify the correlation length $\xi \propto |t|^{-1/y_t}$, so that

$$\boxed{\nu = 1/y_t} \qquad (3.50)$$

At the critical point $t = 0$ we should instead iterate the transformation law (3.48) until $r/b^n = O(r_0)$, where r_0 is some fixed distance, much larger than a or the range of the interaction, so that all the approximations made above are still valid. We then see that $G(r) \propto r^{-2(d-y_h)}$, so that

$$\boxed{\eta = d + 2 - 2y_h} \tag{3.51}$$

The exponents ν and η, related to the spin-spin correlation function, are therefore also given in terms of the relevant renormalization group eigenvalues y_t and y_h. They are thereby also related to the thermodynamic exponents, by scaling relations such as

$$\alpha = 2 - d\nu \tag{3.52}$$
$$\gamma = \nu(2 - \eta). \tag{3.53}$$

Results of this kind clearly require further assumptions than went into the scaling relations such as (3.35, 3.36) for the thermodynamic exponents. A crucial input is that the fixed point hamiltonian be sufficiently short range. For long range interactions (see Section 4.3), for example, (3.53) fails.

(3.52) is an example of a *hyperscaling* relation, since it connects the singularity in the specific heat with the behaviour of a correlation length, which may be inferred, for example, from the spin correlation function. This relation may fail when a dangerous irrelevant variable (see the previous section) influences the scaling form of the free energy but not of the correlation functions. This is what happens above the upper critical dimension (see Section 5.4), and in some problems, even below it. Examples are the random field Ising model (Section 8.4), and the branched polymer problem (Section 9.4).

3.8 Scaling operators and scaling dimensions

The above discussion of the spin-spin correlation function of the Ising model may readily be generalised to an arbitrary correlation function. Near a general fixed point, the scaling variables u_i are linear combinations of the deviations $K_a - K_a^*$ of the original couplings from their fixed point values. Each of these couples to a unique possible interaction term S_a in the hamiltonian. Each S_a may be expressed in terms of the fundamental degrees of freedom of the problem (for the Ising model, the spins $s(r)$). For example, in the Ising model, the set $\{S_a\}$ would be expressible as linear combinations of arbitrary products of spins on different sites. However, the assumption that only short-range couplings are important means that these objects should be *local*, in some sense to be made more precise. It has become common to call these composite objects *operators*. The reason is that, when a suitable continuum limit is taken, the statistical mechanics model is formally identical to a quantum field theory, in which these quantities become operators which may represent observables in the sense of quantum mechanics. However, it is important to realise that in this book, except when so stated explicitly, these 'operators' are commuting quantities.†

Given a complete set of operators S_a we may form suitable linear combinations ϕ_i, called *scaling operators*, coupling uniquely to each of the scaling fields u_i, so that

$$\sum_i u_i \phi_i = \sum_a (K_a - K_a^*) S_a. \qquad (3.54)$$

It is then straightforward to generalise the argument of the previous section to show that, as $|r_1 - r_2| \to \infty$, $\langle \phi_i(r_1)\phi_i(r_2) \rangle \propto |r_1 - r_2|^{-2x_i}$, where

$$\boxed{x_i = d - y_i} \qquad (3.55)$$

This equation, which relates the renormalization group eigenvalue of a scaling variable to the behaviour at the fixed point of the two-point correlation function of the operator to which it couples, is one of the most fundamental and general results of the renormalization group. The quantity x_i is called the *scaling dimension* of the scaling operator ϕ_i. The relation (3.55) may be understood if we assume that it is possible to take the continuum limit of the hamiltonian, in such a way that

$$\sum_i u_i \sum_r \phi_i(r) \to \sum_i u_i \int \phi_i(r) \frac{d^d r}{a^d}, \qquad (3.56)$$

where a^d is the volume of the unit cell. If, under a renormalization group transformation where $a \to ba$ and $u_i \to b^{y_i} u_i$, we demand

---

† An alternative terminology is to call the u_i scaling 'fields', and the ϕ_i scaling 'densities'. This is, however, especially confusing when used in the context of quantum field theory, and we shall avoid it.

that the partition function be invariant, this may then be ensured by requiring that $\phi_i(r) \rightarrow b^{x_i}\phi_i(r)$, with x_i given by (3.55).

As an example, consider the local energy density $E(r)$, which, for the Ising model, is the product $s(r_1)s(r_1')$ of neighbouring spins. It has a scaling dimension $x_E = d - y_t = d - \nu^{-1}$. Therefore, at the critical point, its two-point correlation function decays as

$$\langle E(r_1)E(r_2)\rangle \sim |r_1 - r_2|^{-2d+2/\nu}, \qquad (3.57)$$

which is confirmed by exact results in two dimensions. Actually, in writing results such as (3.57) we should remember that operators like $E(r)$ are not themselves scaling operators, but only linear combinations thereof. Only *scaling* operators have a pure power behaviour for their correlation functions. The lattice energy operator will, in general, be a combination of all possible scaling operators which transform in the same way under the symmetries of the fixed point hamiltonian. The most relevant of these, and hence the one with the smallest scaling dimension, will be the operator ϕ_t, whose 2-point function, at the fixed point, will have the pure power-law form (3.57). In general, for finite separations, (3.57) should be replaced by †

$$\langle E(r_1)E(r_2)\rangle = \sum_{i,j}\frac{A_{ij}}{|r_1 - r_2|^{x_i+x_j}}. \qquad (3.58)$$

Even this is not the full story, since, in general, a system at its critical point is not at the fixed point, but, as discussed in Section 3.6, its hamiltonian differs by irrelevant operators. Just as for the free energy, these lead to correction to scaling terms which now show up as corrections to (3.58) of the form $|r_1 - r_2|^{-x_i-x_j-\sum_k' |y_k|}$, where the sum is over the eigenvalues of some subset of irrelevant scaling fields.

The usefulness of the concept of scaling dimension is not restricted to the two-point correlations. For example if we consider an N-point correlation function, the same kind of arguments that led to (3.55) now imply that this has the homogeneity property

$$\langle \phi_1(r_1)\phi_2(r_2)\ldots\phi_N(r_N)\rangle =$$
$$R^{-x_1-\cdots-x_N}\langle \phi_1(r_1/R)\phi_2(r_2/R)\ldots\phi_N(r_N/R)\rangle, \qquad (3.59)$$

---

† In fact, conformal invariance (see Section 11.2) implies that the terms with $i \neq j$ vanish in this sum.

although, unlike the case $N = 2$, this (together with translational and rotational invariance) is not sufficient to fix its actual form.

3.9 Critical amplitudes

In the previous sections we saw how the various critical exponents are directly related to the renormalization group eigenvalues at the corresponding fixed point. Since this fixed point controls the critical behaviour of all systems in a given universality class, it follows that the critical exponents are universal. However, many other quantities are universal besides the exponents. In this section we focus on the amplitudes which multiply the power law singularities in thermodynamic quantities near the critical point. It is necessary to go back to the inhomogeneous transformation law for the free energy (3.25). Iterating this n times

$$f(\{K\}) = \sum_{j=0}^{n-1} b^{-jd} g(\{K^{(j)}\}) + b^{-nd} f(\{K^{(n)}\}), \qquad (3.60)$$

where $\{K^{(j)}\})$ is the jth iterate of $\{K\}$. Now take the limit $n \to \infty$. Since all initial values of the $\{K\}$ in the same phase ultimately iterate into the same stable fixed point $\{K^{(\infty)}\}$, the second term on the right hand side tends to zero. The first term is a weighted sum of the values of $g(\{K\})$ at points along the renormalization group trajectory which leads from the starting value of $\{K\}$ up to this stable fixed point (see Figure 3.9). For starting points sufficiently close to the critical surface, this trajectory (almost) breaks into two pieces: one which closely approximates a trajectory along the critical surface and which ends up near the critical fixed point; and one starting from near the vicinity of the critical fixed point and ending at the stable fixed point. This latter part closely approximates the *unique* trajectory leading to the stable fixed point which, if the arrows were reversed, would arrive at the critical fixed point. We may call this the outflow trajectory.† The sum in (3.60) may thus be broken into two pieces accordingly. The first contribution may be shown to give only correction to scaling terms

---

† It is the subspace of all models which correspond to renormalizable continuum quantum field theories, where all the irrelevant couplings have been set to zero.

in the free energy (as expected from the discussion in Section 3.6). To evaluate the second term, we may proceed as follows.

Define a coordinate $\tilde{u}_t$ along the outflow trajectory. Close to the critical fixed point, we may take $\tilde{u}_t \sim u_t$, the thermal scaling variable, but, further away, they will deviate from each other due to the curvature of the trajectory. However, in general, we may define $\tilde{u}_t$ so that it is a *nonlinear scaling variable* (see Ex. 3.5), which means that it transforms homogeneously under the renormalization group, $\tilde{u}_t \to b^{y_t} \tilde{u}_t$, for *all* values of $\tilde{u}_t$, not just close to the critical fixed point. Thus, the contribution to the free energy from the outgoing part of the trajectory, which will turn out to give rise to the leading singular behaviour as $t \to 0$, may be written

$$f(t) \sim \sum_{j=0}^{\infty} b^{-jd} g(b^{jy_t} \tilde{u}_t), \qquad (3.61)$$

where $\tilde{u}_t = t/t_0$. Since we are interested in extracting the behaviour as $t \to 0$, it is permissible to replace the sum over j by an integration, and to change to the integration variable $s \equiv b^{jy_t}(t/t_0)$, whence

$$f(t) \sim \frac{\tilde{u}_t^{d/y_t}}{y_t \ln b} \int_{u_t}^{\infty} s^{-d/y_t - 1} g(s) ds. \qquad (3.62)$$

All we need assume about the function $g(s)$ is that it is analytic, as was argued in Section 3.4, and that it and all its derivatives approach zero sufficiently fast as $s \to \infty$.

Superficially the right hand side of (3.62) appears to exhibit the $\tilde{u}_t^{d/y_t} = t^{2-\alpha}$ behaviour expected of the singular part of the free energy, as argued in Section 3.5. However, this conclusion is valid only if it is permissible to set the lower limit of the integration to zero, which is, in general, not the case since $g(0) \neq 0$ and the integral would diverge. It is necessary first to integrate by parts, in order to increase the power of s in the integrand. After p such integrations, the contribution from the lower limit is of the form $\tilde{u}_t^{d/y_t} \tilde{u}_t^{-d/y_t + p - 1} g^{(p-1)}(\tilde{u}_t)$, which is analytic in t. If we integrate by parts a sufficient number of times so that $p > d/y_t$, the remaining integral is of the form†

$$\int_{u_t}^{\infty} s^{-d/y_t+p-1} g^{(p)}(s)\, ds, \qquad (3.63)$$

in which is it now permissible to set the lower limit to zero (the corrections to this may once again be shown to be analytic in t).

The conclusion is that the contribution to the free energy near $t = 0$ from the outgoing part of the trajectory is of the form

$$f(t) \sim A_{>,<}\, |t|^{2-\alpha} + \text{terms analytic in } t, \qquad (3.64)$$

where the subscripts $(>, <)$ on the amplitude A indicate that we expect it to be different for $t > 0$ and $t < 0$. This is because, in each case, the renormalization group trajectory along which g is integrated away from the critical fixed point is different. The amplitudes $A_{>,<}$ consist, apart from trivial factors, of a *universal* integral $\int_0^{\infty} s^{-d/y_t+p-1} g^{(p)}(s) ds$, and the non-universal scale factor $t_0^{-2+\alpha}$. This latter factor arises from the renormalization of u_t which occurs in the first part of the renormalization group flow from the critical theory into the vicinity of the critical fixed point. It is therefore the same no matter what the sign of t. We conclude that, although the individual amplitudes $A_{>,<}$ are not universal, their *ratio* $A_>/A_<$ is.

This is just one example of a universal amplitude ratio. In general, combinations of critical amplitudes in which the non-universal scale factors like $t_0, h_0, \ldots$ do not enter are expected to be universal. These non-universal factors determine how the non-linear scaling variables, which describe the position along the outflow trajectory, are related to the physical 'knobs' which the experimentalist may adjust. Once on the outflow trajectory, however, everything is universal.

Another important example of a universal amplitude combination is the quantity $f_s \xi^d$, where ξ is the correlation length. Since f_s is the singular part of the free energy per unit volume, this combination is the free energy per correlation volume. In terms of the exponents defined earlier, it should scale as $t^{2-\alpha} \cdot t^{-d\nu}$, and so should be independent of t if the hyperscaling relation (3.52) holds. The universality of the numerical value of $f_s \xi^d$ is therefore a stronger statement of hyperscaling.

---

† An interesting situation arises when d/y_t is an integer (see Ex. 3.6).

The above arguments show, in general, that such universal combinations are in principle calculable from the renormalization group, although they depend on properties of the renormalization group flows not just at the critical fixed point, but on more global quantities.

Anisotropic scaling

In deriving (3.48) we assumed that the two-point correlation function depends, at least at large distances, only on the magnitude $r_{12} = |r_1 - r_2|$ of the separation between the two points. As discussed there, this is certainly a correct assumption if the underlying lattice model is sufficiently isotropic. If this is not the case, however, various other possibilities may occur. These may be understood by examining the Fourier transform of the exchange interaction $J(r - r')$, which, for small wave numbers $\mathbf{k} = (k_x, k_y, \ldots)$, has the form

$$\tilde{J}(\mathbf{k}) = J(1 - \sum R_i^2 k_i^2 + O(k^4)). \tag{3.65}$$

If none of the R_i^2 vanishes, we may simply perform a rescaling of the coordinates so that, at least to $O(k^2)$, (3.65) is isotropic. Although this argument is strictly valid only within mean field theory, since, as will be argued in Chapter 4, the higher powers of k are irrelevant at the isotropic fixed point, it in fact continues to hold when the fluctuations are included.[†]

However, a much more severe kind of anisotropy may arise if one of the R_i^2 in (3.65) vanishes. For example, $\tilde{J}$ may have the expansion for low wave number

$$\tilde{J}(k) = J(1 - R^2 k_\perp^2 - \lambda R^4 k_\parallel^4 + \cdots), \tag{3.66}$$

where $\mathbf{k} = (k_\parallel, \mathbf{k}_\perp)$, and λ is dimensionless constant. Such a behaviour arises, for example, at the *Lifshitz point* in a system with competing ferromagnetic nearest neighbour and antiferromagnetic next-nearest neighbour interactions in one particular direction. In this case, it is clearly not possible to remove the anisotropy by a simple rescaling of the coordinates. In mean field

---

† However, the quantitative rescaling required to render the model isotropic does depend on the value of these irrelevant terms, and so is not simply given by the mean field result.

theory, the Fourier transform of the correlation function for such a system has the form

$$\tilde{G}(k) \propto (\lambda k_{\parallel}^4 + k_{\perp}^2 + \xi^{-2})^{-1}. \tag{3.67}$$

However, when the fluctuations are included, there is a further complication in that λ is, in general, renormalized, since there is no symmetry which protects this. Because of the intrinsic anisotropy, it no longer makes sense to rescale all distances, both in the $\perp$ and $\parallel$ subspaces, by the same constant b. Instead, one should rescale only $r_\perp$, for example, and allow λ to be renormalized in such a way that the low wave-number physics is preserved. A little thought then shows that, on dimensional grounds, the renormalization group equation for λ must take the form

$$\lambda' = \lambda f(\{K\}), \tag{3.68}$$

where f depends on all the other dimensionless couplings. At the fixed point, then, we expect $\lambda' = f(\{K^*\})\lambda \equiv b^{-2/z}\lambda$, defining the *anisotropic scaling exponent z*. This shows up in several ways. For example, the scaling form (3.49) is replaced by

$$G(r_{\parallel}, r_\perp, t) = \xi_\perp^{-(d-2+\eta)} \Psi\left(r_\perp/\xi_\perp, r_{\parallel}/r_\perp^z\right), \tag{3.69}$$

where $\xi_\perp \propto |t|^{-\nu_\perp}$ is the correlation length in the $\perp$ directions. This scaling form then implies that the correlation length in the other direction diverges like $|t|^{-\nu_{\parallel}}$, where $\nu_{\parallel} = z\nu_\perp$. The correlation lengths in the two directions therefore exhibit quite different scaling behaviour. Note that, in the mean field approximation, we have $\nu_\perp = \frac{1}{2}$, $\nu_{\parallel} = \frac{1}{4}$ and $z = \frac{1}{2}$ in this example. (3.69) is the general form for a two-point correlation function exhibiting *anisotropic scaling*. Apart from the example of a Lifshitz point, we shall see it in the problem of directed percolation (p.200), and, when the $\parallel$ direction is interpreted as time, in quantum critical behaviour (p.76) and critical dynamics (p.192).

Exercises

3.1 The one-dimensional Ising model in a magnetic field has the reduced hamiltonian $\mathcal{H} = -K \sum_j s_j s_{j+1} - h \sum_j s_j$. By summing over every other spin, show that you can define a renormalization group transformation with $b = 2$. It is useful to express this in terms of the variables $x = e^{-2K}$ and $y = e^h$.

Sketch the flows in the (x, y) plane and indicate the fixed points.

3.2   The one-dimensional three-state Potts model is defined as follows: at each site j is a 'spin' which may take the values $1, 2$ or 3. The reduced hamiltonian is $\mathcal{H} = -K \sum_j \delta_{s_j, s_{j+1}}$. Using the same decimation procedure with $b = 2$ as above, find the renormalization group equation and show that there are no non-trivial fixed points, as expected in one dimension.

3.3   By solving the one-dimensional Ising model exactly (e.g. by the transfer matrix method – see the references by Stanley and by Baxter in the Bibliography) show that the combination $f_s \xi$ is constant in the low temperature limit when $\xi \to \infty$, and calculate its numerical value. Do the same calculation for a spin 1 Ising model, with the same form for the hamiltonian but with spins s_j taking the values $0, \pm 1$, and verify that the amplitude combination is indeed universal.

3.4   There is an amplitude corresponding to each of the critical exponents defined for a ferromagnet on p.7. List as many combinations of these as you can which should be universal according to the arguments of Section 3.7.

3.5   Suppose that the infinitesimal renormalization group equations have the form $dg_i/d\ell = -\beta_i(\{g\})$ where the right hand sides have a perturbative expansion of the form $\beta_i = -y_i g_i + \sum_{jk} c_{ijk} g_j g_k + \cdots$. Show that, in general, it is possible to define, order by order in the gs, *non-linear* scaling variables $g'_i \equiv g_i + \sum_{jk} d_{ijk} g_j g_k + \cdots$ so that the renormalization group equations simplify to $dg'_i/d\ell = y_i g'_i$ exactly, with no higher order terms. Under what circumstances does such a transformation fail (to the order stated)? If there was a non-trivial fixed point at some finite value of the g_i in terms of the old variables, what has happened to it in terms of the non-linear scaling variables?

3.6   Show that when d/y_t is an integer, the free energy has a singularity of the form $|t|^{d/y_t} \ln |t|$. What can you say about the ratio of the amplitudes of this singular behaviour for $t > 0$ and $t < 0$?

3.7   Show that the contribution of the inflow part of the trajectory in the calculation of the free energy using (3.60) leads to the expected correction to scaling terms.

4

Phase diagrams and fixed points

In the previous chapter, the ferromagnetic Ising model provided a simple example of a phase diagram with an associated fixed point structure of the renormalization group flows. There were stable fixed points corresponding to low and high temperature phases, and a critical fixed point controlling the behaviour of critical Ising systems. However, more realistic systems often have more complicated phase diagrams, and therefore a richer fixed point structure. In this chapter we study some of these examples, and show how, even with a rather qualitative description of renormalization group flows, it is possible to understand phase diagrams from the renormalization group viewpoint. More importantly, when more than one non-trivial fixed point is present, the question arises as to which is the dominant one in a particular region of the phase space. The renormalization group answers this question through the theory of *cross-over behaviour*. The existence of such phenomena, whereby different fixed points may influence the properties of the same system on different length scales, is totally absent in mean field treatments.

4.1 Ising model with vacancies

As a first example, consider a generalisation of the Ising model in which the spin variables $s(r)$ may take the value 0 as well as ± 1. This may be viewed as the classical version of a quantum spin-1 magnet, or as a lattice gas of magnetic particles, with $|s(r)|$ playing the role of the occupation number. Alternatively, the sites where $s(r) = 0$ may be thought of as vacancies in the Ising model. Note, however, that they are free to move around: in the language of Section 8.1, they are annealed. The reduced hamiltonian is†

$$\mathcal{H} = -\frac{1}{2}\sum_{r,r'} J(r - r')s(r)s(r') + \Delta\sum_r s(r)^2 - H\sum_r s(r). \quad (4.1)$$

This is known as the Blume–Capel model. The new parameter Δ may be thought of as a chemical potential for the vacancies. For most of this discussion we shall take $H = 0$.

When $\Delta \to -\infty$, the configurations with $s(r) = 0$ are completely suppressed in the partition function, and we recover the usual Ising model. For large negative Δ, it is possible to do perturbation theory in e^{Δ}. The first order correction is due to a single vacancy, and the corresponding change in the free energy is proportional to $\sum_r \langle e^{-\sum_{r'} J(r-r')s(r)s(r')} \rangle$. Near the critical point, this may be expanded in a sum of scaling operators, all of which will be even under spin reversal. The dominant term is therefore proportional to the energy density of the unperturbed Ising model, which couples linearly to the reduced temperature. Therefore the leading effect is that of a slight shift in the critical temperature, without changing the universal critical properties.

It is also simple to analyse the model at zero temperature (which corresponds to all the reduced parameters becoming large in fixed ratios), by minimising $\mathcal{H}$. Because the interaction is ferromagnetic, the only candidates for the ground state have $s(r)$ independent of r. The ordered states with all $s = \pm 1$ have energy per site $\Delta - J$, where as usual $J = \sum_r J(r)$, while the state with all $s = 0$ has zero energy. The latter is therefore the ground state when $\Delta > J$, and there is a zero-temperature transition to an ordered state at $\Delta = J$. This is a first order transition, in the sense that there are discontinuities in the magnetisation and the derivative of the energy with respect to Δ. Also the correlation length is zero at this point, since there are no fluctuations. At finite but low temperatures, this first order transition must persist for some distance into the phase diagram, since the thermodynamic quantities are not singular, and perturbation theory should have a finite radius of convergence. Therefore, if we consider the phase

† It is also possible to add an interaction proportional to $\sum_{r,r'} K(r - r')s(r)^2 s(r')^2$. This adds another interesting dimension to the problem, but will not be considered here.

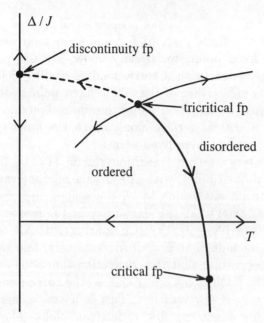

Figure 4.1. Schematic phase diagram of the Ising model with vacancies.

boundary between the ordered and disordered phases, at some point along this boundary the transition must change from being first order to being second order. This point is called a *tricritical point*. The phase diagram of this model is shown schematically in Figure 4.1. In fact, it is straightforward to obtain the main features of this diagram from mean field theory (see Ex. 2.5). The additional term in the hamiltonian allows the coefficient of the M^4 term to change sign in the mean field free energy function $f(M)$, so that the potential changes from the form in Figure 2.1 to that in Figure 2.3. The point at which the coefficients of both the M^2 term and the M^4 term vanish is the mean field approximation to the tricritical point.

Let us now analyse the expected form of the renormalization group flows in this model. Although these take place in an infinite-dimensional space, we may simplify this and consider only the projections of the renormalization group flows in the $(J^{-1}, \Delta/J)$ plane, as shown in Figure 4.1. As usual, there will be stable fixed points which are sinks for the each of the possible phases. There will be a fixed point with the properties of the ordinary critical Ising model at a finite value of J^{-1} and some large negative value of Δ. This fixed point, we already know, has only one relevant thermal eigenvalue, with a corresponding scaling field t describing flows towards either of the stable fixed points, depending on its sign. Thus the direction along the phase boundary must correspond to a critical surface along which the flows go locally in towards the critical Ising fixed point.

The zero temperature transition should also be described by some kind of fixed point, with $\Delta - J$ as a relevant variable there. If we denote its eigenvalue by y, the scaling arguments of Section 3.4 then show that the singularity in the free energy should be of the form $|\Delta - J|^{2-d/y}$. Comparing with the exact result above, we conclude that $y = d$. Alternatively, this result follows from the observation that the correlation functions of the energy density at the fixed point are trivial, so the corresponding scaling dimension $x = d - y$ vanishes. This is a very general property of fixed points describing first order transitions. They are called *discontinuity fixed points* for this reason. A similar kind of fixed point controls the low temperature phase of the ordinary Ising model, with the external magnetic field H playing the same role as $\Delta - J$ in the above example. The fact that H has eigenvalue $y = d$ at this fixed point ensures that there is a discontinuity in the variable conjugate to H, the magnetisation.

At the discontinuity fixed point, the temperature T is irrelevant. This may be seen from the argument used to discuss the zero-temperature fixed point of the Ising model in Section 3.2. In fact, we see from this argument that, at such a discontinuity fixed point, T has eigenvalue $(1 - d)$. Returning to Figure 4.1, we see that there are two attractive fixed points on the critical surface, and therefore their domains of attraction must be separated by another, unstable fixed point. This we may term the tricritical fixed point, since it controls the scaling behaviour in the vicinity of the tricritical point. Although the topology of the renormalization group flows in Figure 4.1 looks very similar to that of the actual phase diagram, it should be stressed that the former is only a projection of the actual flows.

These arguments show that the tricritical fixed point has two relevant thermal eigenvalues, y_{t1} and y_{t2}, rather than the single one at the critical fixed point. One of the corresponding scaling fields, u_{t1} say, will correspond to the flows towards either of the other two fixed points, i.e., along the phase boundary. The corresponding eigenvalue will determine the rate of divergence of the correlation length, which is finite along the first order portion of the phase boundary, as the tricritical point is approached. The other relevant thermal scaling field u_{t2} will, in general, be some linear combination of the reduced chemical potential $\delta = (\Delta - \Delta_c)/\Delta_c$ and the reduced temperature $t = (T - T_c)/T_c$. It turns out that this scaling field is more relevant than the one generating motion along the phase boundary. Thus the leading singularity in the free energy, as the tricritical point is approached along any direction in the phase diagram *except* along the phase boundary, will be dictated by the eigenvalue y_{t2}. If we cross the phase boundary anywhere along its second order portion other than exactly at the tricritical point, the long distance behaviour should be described by the critical, rather than the tricritical, fixed point, and we should observe the corresponding exponents sufficiently close to the critical line. However, close to the tricritical point, there is a cross-over phenomenon, similar to that described for a different example in the next section.

The phase diagram of this model in the presence of a non-zero magnetic field, in the vicinity of the tricritical point, is even more interesting. A mean field calculation suggests that 'wings' sprout out of the first-order part of the phase boundary in the $H = 0$ plane. These are illustrated schematically in Figure 4.2. These wings correspond to surfaces of first order transitions, across which the order parameters $\langle s \rangle$ and $\langle s^2 \rangle$ are discontinuous. Yet there is no symmetry breaking distinguishing the phases on either side, and so we should expect that, at sufficiently high temperature, they should become identical. Thus these wings terminate along the lines of second order transitions shown. A section across a wing is very similar to the phase diagram of a simple fluid shown in Figure 1.3, and therefore we expect the edge of the wing to correspond to a line of second order transitions in the same universality class as the Ising model. There are therefore *three* lines of second order transitions meeting at the tricritical point, which is in fact the origin of its nomenclature.

The three-dimensional projections of the renormalization group

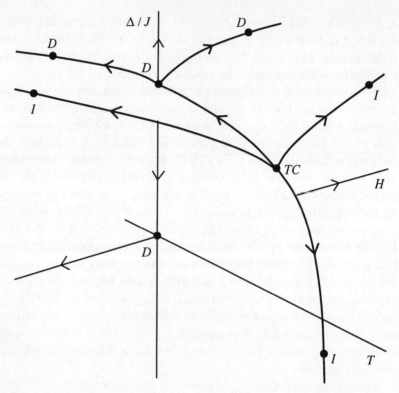

Figure 4.2. Schematic phase diagram near the tricritical point in a magnetic field.

flows responsible for this rich behaviour are also shown in Figure 4.2. As well as the tricritical fixed point, there are various discontinuity fixed points (D), responsible for the various first order transitions, and no less than three distinct critical fixed points (I), all in the same universality class as the Ising model!† This shows that the tricritical point must have (at least) two relevant magnetic scaling variables, one coupling to a generic magnetic perturbation, and the other describing flows along the edges of the wings towards the Ising-like fixed points.

The reader may now well imagine that there are higher multicritical points in the Ising system, corresponding in mean field

---

† This example serves to warn us that two critical systems may be in the same universality class without being described by the same fixed point.

theory to the vanishing of more and more of the coefficients in the expansion of $f(M)$ in powers of M, and described by fixed points with increasing numbers of relevant thermal eigenvalues. This is indeed the case, although such multicritical models become less and less physically relevant. This is because, as discussed in Section 3.3, the number of relevant eigenvalues is the same as the number of 'knobs' the experimentalist has to tune to reach the multicritical point. Since, in any given experimental system, the available number is small, it becomes increasingly unlikely that higher order multicritical points are experimentally accessible.

4.2 Cross-over behaviour

We now discuss another model which more clearly illustrates the phenomenon of cross-over. Consider the space of hamiltonians describing short range ferromagnetic systems in which the degrees of freedom are represented by a three-component vector $\mathbf{s}(r) = (s_x(r), s_y(r), s_z(r))$. Depending on the degree of symmetry of the hamiltonian, the critical behaviour of such a system might be in the universality class of the Heisenberg, XY, or Ising models. This is illustrated by the model hamiltonian

$$\mathcal{H} = -\tfrac{1}{2}\sum_{r,r'} J(r-r')\mathbf{s}(r)\cdot\mathbf{s}(r') - D\sum_r s_z(r)^2. \qquad (4.2)$$

Such a model describes a Heisenberg magnet with *single ion uniaxial anisotropy*. The anisotropic term is uniaxial, that is, it couples preferentially to one axis (in this case, the z-axis), and breaks the $O(3)$ symmetry down to a direct product of XY ($O(2)$) and Ising symmetries. Alternatively, the same symmetry breaking may be due to *exchange anisotropy*, in which the first term in (4.2) is replaced by

$$\sum_{r,r'} \left[J_\perp(r-r')\left(s_x(r)s_x(r') + s_y(r)s_y(r')\right) + J_\parallel(r-r')s_z(r)s_z(r') \right]$$

$$(4.3)$$

Either of these forms might arise if the magnetic ions are arranged in a crystal with uniaxial symmetry. For the purposes of illustration, we shall consider the single ion example of (4.2).

When $D = 0$, the model has the full $O(3)$ symmetry and its critical behaviour is in the Heisenberg universality class, controlled

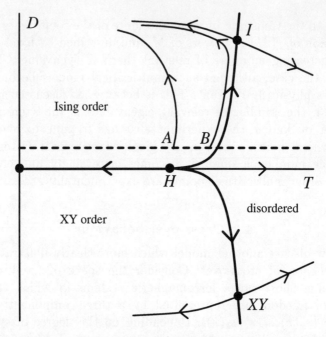

Figure 4.3. Projected RG flows for the Heisenberg model with uni-axial anisotropy.

by the Heisenberg fixed point. When D is large and positive, the Ising-like configurations with $s_z = \pm 1$ are favoured, and we expect the transition to be in the Ising universality class. Similarly, when D is large and negative, the configurations with $|s_z| \ll 1$ are favoured, leading to XY-like critical behaviour. Therefore, the renormalization group flows, projected into the (T, D) plane, will exhibit three critical fixed points, one for each universality class, illustrated in Figure 4.3. In the neighbourhood of the Ising and XY fixed points, the flows along the critical surface will be towards each of them, since they each possess only one relevant thermal scaling variable (which is the one which takes the system away from the critical surface). Although we can make strong statements about the nature of the critical behaviour only for $|D|$ large, the simplest way of connecting up the renormalization group flows, illustrated in Figure 4.3, suggests the phase diagram shown there. Either Ising-like or XY-like critical behaviour should be expected for arbitrarily small, but non-zero, D, depending on

its sign. There is a first order transition across the line $D = 0$ for
$T < T_c$. The point $D = 0$, $T = T_c$, being at the junction of two
lines of critical points, is an example of a *bicritical* point.

For definiteness, suppose $D > 0$ but small. In a typical experiment, we might traverse the dashed line in the phase diagram
as the temperature is varied. Two typical renormalization group
trajectories starting from points on this line are illustrated in Figure 4.3. Starting at point A, not too close to the critical surface,
the renormalization group trajectory rapidly leaves the vicinity of
the Heisenberg fixed point, heading (in this case) towards the low
temperature fixed point. Since the trajectory never approaches
the Ising fixed point, the only critical exponents relevant for the
calculation of the free energy (which, from (3.61), is given by a
weighted sum along this trajectory) will be those of the Heisenberg fixed point. However, if we start at a temperature sufficiently
close to the critical temperature (point B), the trajectory now will
pass very close to Ising fixed point. Since it is a fixed point, the
trajectory will spend a long time in its vicinity, an amount which
actually diverges as the initial point B approaches the critical
surface. The observed critical behaviour should then be related to
that of the Ising model.

To put this on a more quantitative basis, consider the transformation law (3.26) for the singular part of the reduced free energy,
in the vicinity of the Heisenberg fixed point:

$$f_s(t, D) = b^{-nd} f_s(tb^{ny_{tH}}, Db^{ny'_H}), \qquad (4.4)$$

where $y'_H > 0$ is the eigenvalue corresponding to the relevant
variable D.[†] Choosing n in the usual way so that $tb^{ny_{tH}} = O(1)$,
we find the scaling form

$$f_s(t, D) = |t|^{2-\alpha_H} \Psi \left(D|t|^{-y'_H/y_{tH}} \right), \qquad (4.5)$$

where Ψ is scaling function. The quantity $\phi \equiv y'_H/y_{tH}$ is called
the *cross-over exponent*. Note that it is given entirely in terms of
eigenvalues at the unstable Heisenberg fixed point.

When $D = 0$, the specific heat $C \sim \partial^2 f_s/\partial t^2$ behaves, as ex-
pected, like $|t|^{-\alpha_H}$. If D is small, there will be no significant depar-

---

† In writing this we have assumed that D is proportional to a scaling variable.
If we write $s_z^2 \propto \text{const.} + 2s_z^2 - (s_x^2 + s_y^2)$, we see that it transforms according
to a non-trivial irreducible representation of $O(3)$, and therefore cannot
couple to the $O(3)$-symmetric thermal variable t.

ture from this form until we reach reduced temperatures such that $D|t|^{-\phi} \approx 1$, that is, $|t| \approx t_X = D^{1/\phi}$. This is known as the *cross-over temperature*. Note that, if D is very small, we may never reach this regime before other effects such as inhomogeneities due to impurities become important. When $|t| \ll t_X$ the general arguments given above imply that we should observe Ising-like exponents, so that the specific heat $C \propto |t|^{-\alpha_I}$. Although this behaviour does not follow from the perturbative analysis in the vicinity of the Heisenberg fixed point, we may impose it by hand as an additional boundary condition on the scaling function in (4.5). This has interesting consequences for the way that the non-universal features of the ultimate Ising-like critical behaviour depend on D. To see this, rewrite the scaling form for the specific heat as

$$C \sim |t|^{-\alpha_H} \Psi(D|t|^{-\phi}) = D^{-\alpha_H/\phi}(D|t|^{-\phi})^{\alpha_H/\phi}\Psi(D|t|^{-\phi})$$
$$\equiv D^{-\alpha_H/\phi}\widetilde{\Psi}(tD^{-1/\phi}), \qquad (4.6)$$

which defines the new scaling function $\widetilde{\Psi}$. Notice that the only dependence on t is now through this function. We know that C must have a singularity of the form

$$C \sim A(D)\,(t - t_c(D))^{-\alpha_I}, \qquad (4.7)$$

and the only way this can arise is through $\widetilde{\Psi}$ having a singularity

$$\widetilde{\Psi}(tD^{-1/\phi}) \sim a(tD^{-1/\phi} - b)^{-\alpha_I}, \qquad (4.8)$$

where a and b are constants. Therefore

$$C \sim aD^{(\alpha_I - \alpha_H)/\phi}(t - bD^{1/\phi})^{-\alpha_I}. \qquad (4.9)$$

This tells us two things: the dependence $A(D) \propto D^{(\alpha_I - \alpha_H)/\phi}$ of the amplitude of the Ising peak on D (since $\alpha_I > \alpha_H$, this increases with increasing D), and the shift in the critical temperature which gives the *shape* of the phase boundary $t \propto D^{1/\phi}$ for small D. Since, in fact, $\phi < 1$, this boundary has the singular, cusp-like form indicated in Figure 4.3.

The fact that the dependence of the amplitude $A(D)$ of the singular behaviour, and the shape of the phase boundary, two quantities which are measured quite independently, should be simply related to the same cross-over exponent, is a very clear and model-independent prediction of the renormalization group.

4.3 Cross-over to long range behaviour

The cross-over to behaviour characterised by a fixed point of lower symmetry is not the only important type of crossover. As an example, consider the addition of a term $\sum_{r,r'} V(r - r')s(r)s(r')$ to an Ising hamiltonian, where $V(r) \sim v_0 r^{-d-\sigma}$ corresponds to a long range interaction, decaying with a power law parametrised by σ, as $r \to \infty$. A criterion for the relevance of such a term may be established in several ways. For example, consider the first order correction to the free energy from this term, which is

$$\sum_{r,r'} V(r - r')\langle s(r)s(r')\rangle \sim L^d \int V(r)\langle s(r)s(0)\rangle d^d r, \qquad (4.10)$$

where the correlation function is evaluated in the unperturbed system. Substituting in the scaling form (3.49) $\xi^{-d+2-\eta_{SR}}\Phi(r/\xi)$, (where the subscript on η_{SR} is to indicate its value at the short range fixed point), the change in the free energy per site is seen to scale like $\xi^{-(d+\sigma-2+\eta_{SR})}$ as the correlation length $\xi \to \infty$. This is to be compared with the singular part of the unperturbed free energy, which behaves like ξ^{-d} (see Section 3.9.) The long range interaction will therefore be irrelevant if the perturbation is less singular than this, i.e. if

$$\sigma > 2 + \eta_{SR}. \qquad (4.11)$$

Another instructive way of deriving this result is to compute the scaling dimension of the perturbation through its two-point function at the short range fixed point. Denoting $q(R) \equiv \int V(r)s(R + \frac{1}{2}r)s(R - \frac{1}{2}r)d^d r$, we have

$$\langle q(R_1)q(R_2)\rangle \sim \int \int V(r_1)V(r_2)\langle s(R_1 + \frac{1}{2}r_1)s(R_1 - \frac{1}{2}r_1)$$

$$s(R_2 + \frac{1}{2}r_2)s(R_2 - \frac{1}{2}r_2)\rangle d^d r_1 d^d r_2. \quad (4.12)$$

Fortunately, we do not need to know the precise form of the four-point correlation function in (4.12): by scaling the integral must behave at the fixed point as $R_{12}^{2d-2(d+\sigma)-4x_s}$, where x_s is the scaling dimension of the magnetisation operator s. This implies that the scaling dimension of q is $x_q = 2x_s + \sigma$, and it is irrelevant when $x_q < d$. This recovers the previous result (4.11), using the scaling relations of Section 3.8.

Therefore, even power law interactions, if they decay sufficiently rapidly, do not destabilise the short range fixed point. The criterion (4.11) gives, *a posteriori*, a more precise condition for the various consequences following from the assumption of a short-range fixed point hamiltonian to be applicable. When $\sigma < 2 + \eta_{\mathrm{SR}}$, we expect to see a cross-over to a new, long range fixed point. In fact, there is a separate fixed point for each value of σ.

The existence of such cross-overs indicates that the critical behaviour of a real system can be far richer than that suggested by mean field theory. Ideally, the terms in the hamiltonian may be decomposed according to the symmetries and ranges of their interactions, and arranged in order of decreasing magnitude:

$$\mathcal{H} = \mathcal{H}_0 + \lambda_1 \mathcal{H}_1 + \lambda_2 \mathcal{H}_2 + \cdots, \qquad (4.13)$$

where $1 \gg \lambda_1 \gg \lambda_2 \gg \ldots$ In that case, depending on whether or not the interactions in $\mathcal{H}_{j+1}$ are relevant at the fixed point controlling $\sum_{i=0}^{j} \lambda_i \mathcal{H}_i$, we may see a whole series of cross-overs as the critical point is approached. Of course, in a real system, such a clear separation of length scales is unlikely to occur, and multiple crossovers may then take place simultaneously.

4.4 Finite-size scaling

A slightly different kind of cross-over phenomenon may arise as a result of the geometrical limitations of a given system, rather than its interactions. Consider, for simplicity, a system whose critical behaviour is described by an isotropic fixed point, and whose overall size is determined by the linear dimension L. It is important to realise that the system itself may be finite (for example, a cube of size $L \times L \times L$) or infinite (for example, an infinite slab of thickness L). L, of course, has the dimensions of length, but it will enter into physical quantities such as the free energy per site through the dimensionless variable $N = L/a$, where a is the microscopic distance, for example the lattice spacing. Under renormalization, we rescale $a \to ba$, keeping L *fixed*. (If we were simultaneously to rescale $L \to bL$, we would have changed only the markings on our ruler, not the physical system.) Thus, under the renormalization group, $N \to b^{-1}N$, or, equivalently, $N^{-1} \to bN^{-1}$. It is more useful to think of the free energy and other quantities as depending on the size through N^{-1}, since then the thermodynamic limit corresponds to $N^{-1} \to 0$. The singular part of the free energy thus transforms according to the generalisation of (3.26)

$$f_s(\{K\}, N^{-1}) = b^{-d} f_s(\{K'\}, bN^{-1}), \qquad (4.14)$$

showing that we may think of N^{-1} as a *relevant* scaling variable, with eigenvalue $y = 1$. An important assumption in deriving (4.14) is that a non-zero N^{-1} does not affect the renormalization group equations for the interactions $\{K\}$. This is reasonable as long as the fixed point hamiltonian is short range, but it clearly needs to be examined more carefully in the case of long range interactions.

To illustrate the implications of (4.14), consider the zero-field susceptibility $\chi \sim \partial^2 f_s / \partial h^2$. Iterating (4.14) in the usual way to the point where $t b^{n y_t} = O(1)$ yields the scaling form

$$\chi(t, N^{-1}) \sim |t|^{-\gamma} \phi(N^{-1}|t|^{-\nu}). \qquad (4.15)$$

The argument of the scaling function may be written in a more revealing way: up to factors of order unity, it is simply ξ/N, where ξ is the correlation length of the infinite system (in units of a). Thus, for $N \gg \xi$ we find, as expected, that the behaviour of χ is similar to that of the thermodynamic limit. However, as t decreases, there is a cross-over temperature $|t| = t_X \sim N^{-1/\nu}$ at which finite-size effects become pronounced. What happens for smaller values of $|t|/t_X$ now depends on the geometry. In a truly finite system, for example, the cube mentioned above, there can be no thermodynamic singularities, and therefore the peak in the susceptibility must be rounded off. The extent of this rounding is given by cross-over theory. The scaling form (4.15) may be written equivalently

$$\chi(t, N^{-1}) \sim N^{\gamma/\nu} \tilde{\psi}(t N^{1/\nu}), \qquad (4.16)$$

and, for the truly finite system, the scaling function $\tilde{\psi}$ must be an analytic function of its argument. It is still expected to have a peak, but this will be rounded, and, moreover, its maximum will not necessarily be at $t N^{1/\nu} = 0$, see Figure 4.4. This scaling form then yields two predictions: the *shift* in the effective critical point (as defined by the maximum of χ), which should scale like $N^{-1/\nu}$; and the height of the peak, which is predicted to behave as $N^{\gamma/\nu}$. The direction of shift of the peak depends on the boundary conditions. Periodic boundary conditions act to suppress the

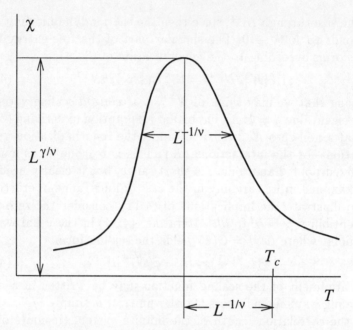

Figure 4.4. Schematic form of the susceptibility in a finite-size system.

effects of fluctuations, since their wave vectors are restricted to be multiples of $2\pi/L$, and therefore to increase the effective critical temperature. Free boundary conditions, on the other hand, allow the degrees of freedom near the boundary to fluctuate more freely, and thus act to depress the effective critical temperature. Fixing the boundary degrees of freedom into an ordered state, however, enhances the order and therefore increases the effective T_c. This illustrates that the scaling function $\tilde\psi$ in (4.16) depends on the boundary conditions as well as the shape of the geometry.

On the other hand, the effective dimensionality of the geometry may be non-zero. For example, consider a prism $L \times L \times \infty$. As $N^{-1} = a/L \to 0$, the rescaled system will exhibit a quasi-one-dimensional behaviour, and the scaling function in (4.16) may be estimated, for $t \ll t_X$, from the solution to the simple one-dimensional model. Since there is no transition in such a model, the finite-size susceptibility will once again be rounded, but in a different manner than that for a system with zero effective dimensionality. It is interesting to consider the behaviour of the correlation length, along the infinite direction, in such quasi-one-dimensional geometries. This transforms under the renormalization group according to

$$\xi(\{K\}, N^{-1}) = b\xi(\{K'\}, bN^{-1}), \qquad (4.17)$$

so that, at the critical point of the *infinite* system $\xi(N^{-1}) \propto N$. Since this correlation length may often be estimated numerically rather simply from the largest eigenvalues of the transfer matrix acting along the infinite direction, the above result gives a powerful way of determining the critical point of the infinite system. (For many two-dimensional systems, the amplitude which is the coefficient of proportionality is related to the scaling dimensions: see Section 11.2.) With a little more work, it is also possible to extract the critical exponents from the knowledge of ξ (see Ex. 4.3).

A third possibility is that the effective dimensionality of the geometry is above the lower critical dimension d_l of the system in question (see Section 6.1.) For example, an Ising-like system in a slab ($L \times \infty \times \infty$) geometry is effectively two-dimensional, and therefore is still expected to have a sharp critical point at some finite temperature. Once again cross-over theory may be brought to bear: in this case the scaling function $\tilde{\psi}$ in (4.16) should have a singularity of the form $(tN^{1/\nu} - b)^{-\gamma_2}$, where γ_2 is the two-dimensional susceptibility exponent, so that χ will have a two-dimensional Ising-like singularity, but at a temperature shifted, once again, by an amount $O(N^{-1/\nu})$, and with an amplitude which scales like $N^{(\gamma_3 - \gamma_2)/\nu}$.

Finite-size scaling is by now well-established theoretically, at least for systems with short range interactions and with no dangerous irrelevant variables. This latter condition is important: for example, above the upper critical dimension $d_c = 4$ the susceptibility at the Ising critical point in a finite cube scales like $L^{d/2}$ rather than $L^{\gamma/\nu} = L^2$ (see Ex. 4.6.) The experimental observability of finite-size scaling is hampered by the fact that nature does not provide us with systems with periodic boundary conditions, and that, for moderate L, the corrections to the bulk thermodynamic behaviour tend to be dominated by the boundary effects described in Chapter 7, rather than by true finite-size effects.

4.5 Quantum critical behaviour

Throughout this book we have, largely for reasons of simplicity, considered critical behaviour occurring in systems described by classical statistical mechanics. Usually, this is permissible, since at finite temperatures the thermal fluctuations dominate the quantum effects. However, this is clearly not always the case, especially in almost degenerate electronic systems, or in low dimensional problems where the classical fluctuations may suppress the critical point to a temperature where quantum effects do become important.

The partition function $Z = \text{Tr} e^{-\beta \widehat{\mathcal{H}}}$ of a quantum statistical system is, of course, intrinsically harder to evaluate because the quantum hamiltonian $\widehat{\mathcal{H}}$ depends on non-commuting degrees of freedom, and we emphasise this with the caret. For this reason, all of the renormalization group methods we have described up to this point fail at the first step. However, progress is possible if one realises that the partition function for many quantum systems may be written as a Feynman path integral in imaginary time τ, schematically written as

$$Z = \int [dpdq] \, e^{-(1/\hbar) \int_0^{\beta\hbar} L(\{p\},\{q\})d\tau}, \qquad (4.18)$$

where L is the Lagrangian, and the functional integral is over the degrees of freedom $\{q\}$ and their canonical conjugates $\{p\}$. The boundary conditions in the imaginary time direction are periodic (respectively, antiperiodic) for bosonic (fermionic) degrees of freedom. Since the Lagrangian L itself usually has the form of an integral (or a lattice sum) over a local Lagrangian density, we may, at least formally, regard (4.18) as the partition function for a *classical* system in $d+1$ dimensions, with classical hamiltonian $\int L d\tau$, in a slab geometry, with thickness $\beta\hbar$. That is, we regard the imaginary time τ as another spatial dimension. Although the above analogy was based on the Feynman path integral formulation, appropriate to quantum many-particle systems, it is in fact completely general. To see this, consider approximating the partition function as the limit

$$Z = \text{Tr} e^{-\beta\widehat{\mathcal{H}}} = \lim_{\Delta\tau \to 0} \left(1 - \Delta\tau\widehat{\mathcal{H}}\right)^{\beta/\Delta\tau}, \qquad (4.19)$$

which is called the Trotter formula. Each factor of $(1 - \Delta\tau\widehat{\mathcal{H}})$ may now be thought of as the transfer matrix for a $(d+1)$-dimensional classical system, acting from one d-dimensional slice to the next. From this point of view, Z is just the classical partition function of this system. To illustrate this with a well-known example, consider a quantum lattice system with a spin-$\frac{1}{2}$ degree of freedom at each site. For simplicity we take the lattice to be one-dimensional. The quantum hamiltonian is

$$\widehat{\mathcal{H}} = -h\sum_j \sigma_j^x - J\sum_j \sigma_j^z \sigma_{j+1}^z, \qquad (4.20)$$

where σ_j^x and σ_j^z are the usual Pauli spin matrices. The model described by this hamiltonian is called the Ising model in a transverse magnetic field. To first order in $\Delta\tau$, we may write

$$1 - \Delta\tau\widehat{\mathcal{H}} \approx \prod_j \left(1 + h\Delta\tau\sigma_j^x\right) \cdot e^{K_1\sum_j \sigma_j^z \sigma_{j+1}^z}, \qquad (4.21)$$

where $\tanh K_1 = J\Delta\tau$. This may be thought of as the row-to-row transfer matrix of a two-dimensional classical Ising model with degrees of freedom $\sigma_{j,i}^z$ and (reduced) exchange couplings K_1 and K_2 in the j and i directions, and classical hamiltonian

$$\mathcal{H} = -\sum_{j,i} \left(K_1\sigma_{j,i}^z \sigma_{j+1,i}^z + K_2\sigma_{j,i}^z \sigma_{j,i+1}^z\right). \qquad (4.22)$$

Since $\Delta\tau \ll 1$, this correspondence is valid only in the limit $K_2 \gg 1$, when the probability of a given spin flipping from one row to the next (which, according to (4.21) is $h\Delta\tau$) is very small. Since this probability is e^{-2K_2} we thus identify

$$K_1 \sim J\Delta\tau \qquad (4.23)$$

$$e^{-2K_2} \sim h\Delta\tau. \qquad (4.24)$$

The corresponding classical system is thus very anisotropic. But, as discussed on p.58, such anisotropy, at least for Ising-like systems, does not alter the critical behaviour. In fact, it is known that the anisotropic Ising model is critical in the thermodynamic limit along the curve in the (K_1, K_2) plane given by $\sinh 2K_1 \sinh 2K_2 = 1$, which translates in the above limit to $h = J$.

The classical two-dimensional Ising model, infinite in both directions, thus corresponds to the one-dimensional transverse Ising model at zero temperature. The correspondence implies that the latter model has a phase transition at zero temperature as a function of the ratio of its coupling constants h/J. The two phases mirror those of the classical Ising model: when $h/J < 1$, the s_j^z degrees of freedom order, and the symmetry under $s_j^z \to -s_j^z$ is spontaneously broken, while $h/J > 1$ corresponds to the high temperature phase of the classical model.

Although the above example used some specific properties of the Ising model, the overall picture is very general. A quantum statistical system in d dimensions corresponds to a classical system in $d+1$ dimensions. The case when the classical system is infinite in all directions corresponds to zero temperature in the quantum system, and phase transitions which may occur as the temperature is varied in the classical system correspond to zero temperature phase transitions as some coupling constant is varied in the quantum system. The temperature β^{-1} of the quantum system is *not* the same as the temperature of the corresponding classical system: instead β corresponds to the finite size in the extra direction. The cross-over theory of finite-size scaling may now be used to discuss the phase diagram of the quantum system as a function of its coupling constants and its temperature. As an example we shall once again consider the Ising model in a transverse field, but for arbitrary d. The mapping to an anisotropic $(d+1)$-dimensional classical Ising model, with exchange couplings K_1 in the d space directions and K_2 in the imaginary time direction, goes through as before. Although no exact information is available on the position of the phase boundary for $d > 1$, we expect it to scale for $\Delta\tau \to 0$ as in $d = 1$, leading to a critical point at zero temperature at some $h/J = (h/J)_c$. Finite temperature T now corresponds to a classical system in a slab of thickness $N = (T\Delta\tau)^{-1}$; thus T is a relevant variable with eigenvalue $y = 1$. For $T > 0$ we therefore expect the behaviour to cross over to that characteristic of a classical d-dimensional Ising system. For $d > 1$ this will have a critical point at some value $K_1 = K_c$, independent of the value of K_2 if the slab is very thin. This latter condition corresponds to taking $T \sim (\Delta\tau)^{-1}$, so that, from (4.23), $K_1 \sim J/T$. Thus the fixed point to which the points on the critical surface flow for $T > 0$ occurs at a particular value of J/T, independent of the value of h.

It is convenient to picture the consequent renormalization group

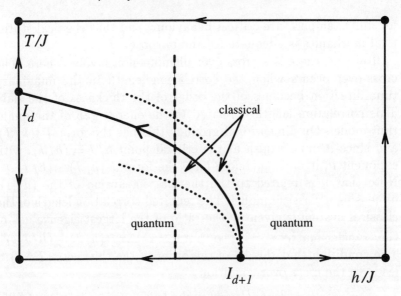

Figure 4.5. Phase diagram and RG flows in the Ising model in a transverse field at finite temperature.

flows as projected onto the $(h/J, T/J)$ plane, illustrated in Figure 4.5. There are two fixed points on the critical surface: an unstable one at $T = 0$, with the critical exponents characteristic of the $(d+1)$-dimensional Ising model, and a stable one at $h/J = 0$, with d-dimensional exponents. As well as these fixed points there are several others of interest. The axis $h/J = 0$ corresponds to rigidly coupled d-dimensional layers, and therefore contains the usual high and low temperature fixed points of the d-dimensional Ising system. The region where $h/J \gg 1$, on the other hand, corresponds quantum mechanically to decoupled Ising spins at finite temperature, or classically to decoupled Ising chains of finite length. From either point of view, we may sketch the renormalization group flows interpolating between all these fixed points as shown in Figure 4.5. This shows that, as long as h/J is less than its critical value, so that the zero temperature state is ordered, the quantum system undergoes a phase transition, as the temperature is varied along the dashed line, which is in the same universality class as the *classical d*-dimensional model. This confirms our presupposition that quantum effects are not important for finite temperature critical behaviour, and that classical statistical mechanics is adequate for the purpose.

However, there is a cross-over phenomenon involved here. The cross-over occurs when the correlation length in the imaginary time direction becomes of the order of the thickness of the slab. This correlation length is related to the energy gap of the quantum model. On dimensional grounds this has the form $J f(h/J)$, and since it must vanish at the critical point $h/J = (h/J)_c$ with exponent ν, it must therefore have the form $J((h/J) - (h/J)_c)^{\nu}$. Note that ν is related to the thermal eigenvalue of the $(d+1)$-dimensional classical model. The required correlation length of the classical system is given by the ratio of the largest eigenvalues of the transfer matrix $e^{-\Delta\tau\widehat{\mathcal{H}}}$, so that $\xi_{\tau}^{-1} \sim \Delta\tau J((h/J) - (h/J)_c)^{\nu}$. The condition that this be of the order of the thickness $(T\Delta\tau)^{-1}$ leads to the cross-over condition

$$(T/J)\,((h/J) - (h/J)_c)^{-\nu} \sim 1. \tag{4.25}$$

Typical curves along which the left hand side is constant are shown by dotted lines in Figure 4.5. Only within this ever-narrowing wedge is classical critical behaviour observed: the region outside the wedge is, for low temperatures, dominated by quantum effects.

The general picture we have arrived at for the phase diagram of the Ising model in a transverse field holds more generally for quantum systems which have a zero temperature critical point. There is an important qualification, however. In the Ising case we were able to argue that the quantum problem mapped, through the transfer matrix, onto an anisotropic classical Ising model. For this model, it is known that any anisotropy in the exchange couplings may be removed by simple relative rescaling of the coordinates (in this case, space and imaginary time), as discussed on p.58. As a result, the correlation lengths in the different directions diverge at the critical point with the same exponent ν, albeit with different amplitudes. However, this property of the corresponding classical model is by no means guaranteed. Since the imaginary time and space directions enter into the problem on very different footings, we might expect the corresponding classical system to exhibit intrinsically anisotropic scaling. The correlation length in the spatial directions may then diverge with an exponent $\nu_{\perp}$ which is not equal to the $\nu_{\parallel}$ which characterises the divergence of the correlation length in the imaginary time direction. The ratio $z \equiv \nu_\parallel / \nu_\perp$ of these exponents is the *dynamic* exponent of the quantum problem (similar to that which will appear in the context of critical dynamics in Chapter 10). Under these circumstances, the above argument for the cross-over behaviour must be generalised, and it is, in fact, straightforward to see that the exponent ν in (4.25) should be replaced by $\nu_\parallel$.

Exercises

4.1 Suppose that the relevant thermal eigenvalues at the tricritical fixed point discussed in Section 4.1 are y_1 and y_2, where the latter corresponds to flows along the critical surface. Assuming that $y_1 > y_2$, what is the form of the leading singularity and the first few corrections to scaling in the specific heat, as the temperature t is varied at fixed Δ in such a way as to pass through the tricritical point? Across the first-order part of the phase boundary, the correlation length and the discontinuity in the magnetisation are finite. How do these quantities behave as the tricritical point is approached along the phase boundary?

4.2 By solving the one-dimensional Ising model exactly, show that the susceptibility in a finite chain of length L has the scaling form (cf. (4.15)) $\chi \sim \xi^{\gamma/\nu} f(\xi/L)$, where ξ is the correlation length of the infinite chain. Calculate the scaling function f for the cases of free and periodic boundary conditions, and show that they are not equal.

4.3 Consider an infinitely long two-dimensional strip of finite width L, and prescribed boundary conditions. Write down a finite-size scaling form for the effective correlation length $\xi(L, t)$ along the strip, in terms of the deviation $t = (T - T_c)/T_c$ of the system from the critical point T_c of the infinite system. Discuss how comparison of numerical results on this quantity for different values of L and T might be used in order to estimate the critical exponent ν, in a case where T_c is not known exactly.

4.4 Consider a three-dimensional Ising model in a slab geometry, with finite thickness L, but effectively infinite in the other two directions. The ultimate critical behaviour of this system should be two-dimensional, with a specific heat singularity of the form $A(L)\ln|T - T_c(L)|$ (see Ex. 3.6). How should the amplitude A depend on L for large L?

4.5 As discussed in Section 8.4, the addition of random fields to the two-dimensional Ising model destroys the long range order and therefore rounds the logarithmic peak in the specific heat. The cross-over exponent is γ. How does the height of the peak depend on the strength of the random field? [Note: since the logarithmic peak in the absence of the random field arises from the mechanism which is the subject of Ex. 3.6, it is necessary to modify that argument in order to expose the required effect.]

4.6 Consider the continuous spin Ising model of Section 5.3 in a hypercube of size L and volume L^d. Write $S(r)$ in terms of its Fourier decomposition $S(r) = S_0 + \tilde{S}(r)$, where S_0 is constant and $\tilde{S}$ contains all the modes with non-zero wave number. In the approximation in which $\tilde{S}$ is neglected, show that the susceptibility has the scaling form $\chi \sim t^{-1}f(L^d/u\xi^4)$, so that finite-size scaling is violated. [Harder] Show that the neglect of $\tilde{S}$ is justified, in considering the dominant singular behaviour, when $d > 4$.

5

The perturbative renormalization group

In this chapter we shall explore the consequences of a system possessing two renormalization group fixed points, and a parameter ϵ (for example, related to the number of dimensions of space) such that, as its value is varied, the fixed points actually collide. When the two fixed points are sufficiently close it is then possible to deduce universal properties at one fixed point in terms of those at the other. Such an analysis is the basis of the ϵ-expansion and many other similar techniques. It also allows us to describe the properties of fixed points with exactly marginal scaling variables.

To see how this works, let us consider a simple example with a single coupling u. For simplicity, we consider an infinitesimal renormalization group (see p.47). Without loss of generality, we may assume that the fixed points are at $u = 0$ and $u = u^*$. Since the beta function is supposed to be analytic in u, we may assume that, for sufficiently small u, its form is determined by its zeroes, and the renormalization group equation therefore has the form

$$du/d\ell = -ku(u - u^*), \tag{5.1}$$

where k is some constant of order unity. Notice that the renormalization group eigenvalues at $u = 0$ and $u = u^*$ are ku^* and $-ku^*$ respectively, and that they are both $O(\epsilon)$ if $u^* = O(\epsilon)$. If $ku^* > 0$, the fixed point at $u = 0$ will be unstable, and, for $u > 0$, the flow will be towards the non-trivial fixed point at $u = u^*$. If we now have some other coupling u_1, say, which is a scaling variable at the $u = 0$ fixed point, in this simple example its renormalization group equation will have the form for small u and u_1

$$du_1/d\ell = y_1 u_1 + b' u u_1 + \cdots, \tag{5.2}$$

where b' is some constant. Therefore the renormalization group eigenvalue of u_1 at the non-trivial fixed point is, to this order, simply $y_1 + bu'^*$, that is, it is changed by an amount $O(\epsilon)$.

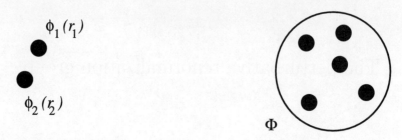

Figure 5.1. The operator product expansion.

Although very simple, this idea is the basis of the ϵ-expansion approach which gives a systematic approximation scheme for the computation of critical exponents. As long as sufficient information is known about the 'trivial' fixed point, all the properties of the non-trivial fixed point may be obtained as an expansion in ϵ. In this book, however, we shall concentrate on only the first order corrections, since they may be obtained rather directly without recourse to much formal development. However, their calculation does depend on a fundamental property of fixed point correlation functions which has not yet been described.

5.1 The operator product expansion

We discussed in Section 3.8 how the correlation functions at a fixed point could be described in terms of scaling operators $\phi_i(r)$, whose two-point correlations have a pure power behaviour. Consider now a correlation function

$$\langle \phi_i(r_1)\phi_j(r_2)\,\Phi \rangle, \tag{5.3}$$

where Φ denotes some arbitrary product of other operators $\prod_l \phi_l(r_l)$. We are interested in how this correlation function behaves in the limit when $|r_1 - r_2|$ is much smaller than the separations $|r_1 - r_l|$ with $l > 2$, as illustrated in Figure 5.1. The idea of the operator product expansion is that, in this limit, (5.3) may be replaced by a sum of the form

$$\sum_k C_{ijk}(r_1 - r_2)\langle \phi_k((r_1 + r_2)/2)\,\Phi \rangle, \tag{5.4}$$

where the sum is over all scaling operators ϕ_k. To see why this should be true, consider replacing ϕ_i and ϕ_j by suitable linear combinations of 'bare' operators S_a, as discussed in Section 3.8. Then the product $\phi_i\phi_j$ may also be expressed in terms of this set, since, viewed from very far away, this is still a local quantity. Each of these may then be re-expressed as a linear combination of the ϕ_k. This argument, although appealing, is at best heuristic. It is therefore comforting to know that the operator product expansion has been proved to be valid to all orders in perturbation theory in simple models. For exactly solvable models, like the two-dimensional Ising model, its validity has also been demonstrated.

The important point about this result is that the functions $C_{ijk}(r_1 - r_2)$ in (5.4) are *independent* of Φ. For that reason, the operator product expansion may be written

$$\phi_i(r_1)\phi_j(r_2) = \sum_k C_{ijk}(r_1 - r_2)\,\phi_k((r_1 + r_2)/2). \qquad (5.5)$$

However, it must be remembered that the above equality is only valid in a 'weak' sense, that is, when both sides of the equation are inserted into correlation functions $\langle \ldots \Phi \rangle$ with all the other operators in Φ much further away. It should also be remarked that the choice of the argument of ϕ_k as the midpoint of r_1 and r_2 is motivated only by considerations of symmetry, and is in fact arbitrary. It is, for example, quite permissible to take the argument of ϕ_k to be r_1. The difference between these two cases involves the Taylor expansion of ϕ_k about r_1. Therefore, to be correct, in the sum over k on the right hand side we should include not just the scaling operators ϕ_k but all possible derivatives of them. These derivative operators would not enter, coupled to scaling fields, in the hamiltonian, since they would give zero on summing over the whole system, but they do enter into the operator product expansion. Fortunately, this subtlety will not trouble us in deriving the renormalization group equations, at least to the order required. Because ϕ_i, ϕ_j and ϕ_k are scaling operators, the form of the function $C_{ijk}(r_1 - r_2)$ is in fact completely determined† by the

† In fact, (5.6) is valid only when all the operators transform as scalars under rotations. The appropriate generalisation to non-scalars is straightforward but very cumbersome, at least for dimension $d > 2$. Fortunately, the most

homogeneity relation (3.59).

$$C_{ijk}(r_1 - r_2) = \frac{c_{ijk}}{|r_1 - r_2|^{x_i + x_j - x_k}}. \tag{5.6}$$

The quantities c_{ijk} are pure numbers called the *operator product expansion coefficients*. Their numerical value depends on the way the operators are normalised: if we fix this by requiring that $\langle \phi_i(r_1)\phi_i(r_2)\rangle = |r_1 - r_2|^{-2x_i}$, for example, the c_{ijk} are then *universal*. As we shall see in the next section, it is these quantities which determine the first order corrections to the renormalization group equations.

5.2 The perturbative renormalization group

Following the discussion in the introduction to this chapter, let us therefore consider a fixed point hamiltonian $\mathcal{H}^*$ which is perturbed by a number of scaling fields, so that the partition function is

$$Z = \mathrm{Tr}\, e^{-\mathcal{H}^* - \sum_i g_i \sum_r a^{x_i} \phi_i(r)}. \tag{5.7}$$

The factors of a^{x_i} are needed because the coupling constants g_i, being linear combinations of the K_a on the lattice, are dimensionless, and the scaling operators have dimension $(\text{length})^{-x_i}$ if we wish to normalise them as above. Let us expand this in powers of the couplings g_i:

$$
\begin{aligned}
Z = Z^* \Bigg[1 &- \sum_i g_i \int \langle \phi_i(r)\rangle \frac{d^d r}{a^{d-x_i}} \\
&+ \tfrac{1}{2} \sum_{ij} g_i g_j \int \langle \phi_i(r_1)\phi_j(r_2)\rangle \frac{d^d r_1 d^d r_2}{a^{2d - x_i - x_j}} \\
&- \tfrac{1}{3!} \sum_{ijk} g_i g_j g_k \int \langle \phi_i(r_1)\phi_j(r_2)\phi_k(r_3)\rangle \frac{d^d r_1 d^d r_2 d^d r_3}{a^{3d - x_i - x_j - x_k}} + \cdots \Bigg],
\end{aligned}
\tag{5.8}
$$

where all correlation functions are to be evaluated with respect to the fixed point hamiltonian $\mathcal{H}^*$. A number of comments should be made at this point. First, we have adopted a continuum notation in which $\sum_r$ is replaced by $\int (d^d r / a^d)$, where a is the microscopic short distance cut-off (for example, the lattice spacing). This is

---

relevant operators are scalars, so we may ignore this complication.

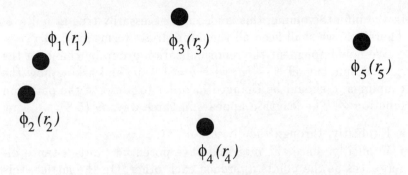

Figure 5.2. Gas picture representing a typical term in (5.7).

justified since the details of the lattice are supposed to be irrelevant. Second, (5.8) purports to represent the power series expansion about a critical point, where it known to be singular. Such an expansion should not exist, and this is made manifest by the fact that the integrals are in fact divergent in the infinite volume limit. The solution to this problem is to first place the system in a finite box, of linear size L. All integrals are then finite, and in fact Z is now non-singular because thermodynamic singularities emerge only in the thermodynamic limit. The drawback to this is that the correlation functions are sensitive to this infrared cut-off, and, for example, the 2-point functions will in general not have a simple power-law form. However, it will turn out that in order to implement the renormalization group, all we need are the short-distance singularities, given by the operator product expansion, which is independent of any boundary conditions.

The expansion in (5.8) may be thought of as a kind of low-density expansion of a gas in the grand canonical ensemble (see Figure 5.2). The gas contains particles of different species, labelled by i, each with its own fugacity g_i. The partition function is given by a sum of one-body terms, two-body terms, and so on. The analogue of the Boltzmann weight of a given configuration is the appropriate correlation function. Note that, although it is always possible to redefine our operators by $\phi_i \to \phi_i - \langle \phi_i \rangle$, so that they have zero expectation value in the infinite volume limit, this will not necessarily be true in a finite box. Similarly, although as discussed in Section 11.2, the 2-point functions vanish when $i \neq j$

in the infinite volume, this is also not necessarily true at finite L. Therefore, we shall keep all the appropriate terms in (5.8).

We now implement the renormalization group by changing the microscopic cut-off $a \to ba$, with $b = 1 + \delta\ell$, and asking how the couplings g_i should be changed in order to preserve the partition function Z. The length a appears in three ways in (5.8):

• Explicitly, through the divisors a^{d-x_i};
• Implicitly, since the integrals have potential short-distance divergences as the points approach each other. On the lattice, this would be regulated by the lattice itself. In the continuum approach, it is more practical to introduce a rotationally invariant cut-off, and to insist that all integrals should be restricted to $|r_i - r_j| > a$. In the interacting gas picture, this corresponds to introducing a hard core repulsion of radius a between the particles. This cut-off, although crude, is quite sufficient for the first order calculation of the renormalization group functions.
• Through the dependence on the system size L, in the dimensionless ratio L/a.

If we choose an infinitesimal rescaling of a, with $\delta\ell \ll 1$, the effects of these different kinds of dependence may be considered separately, as they will contribute additively to the beta functions. The first dependence is simple: the rescaling is compensated in every term in (5.8) by a rescaling

$$g_i \to (1 + \delta\ell)^{d-x_i} g_i \sim g_i + (d - x_i)g_i\delta\ell. \qquad (5.9)$$

The effect of the change in the cut-off may be evaluated by use of the operator product expansion (5.5). To see this explicitly, consider the simplest case, of the second order term in (5.8). After changing $a \to a(1 + \delta l)$, we may break up the integral as

$$\int_{|r_1-r_2|>a(1+\delta\ell)} = \int_{|r_1-r_2|>a} - \int_{a(1+\delta\ell)>|r_1-r_2|>a} . \qquad (5.10)$$

The first term simply gives back the original contribution to Z. The second term may be expressed using the operator product expansion, as

$$\frac{1}{2}\sum_{ij}\sum_{k} c_{ijk}a^{x_k-x_i-x_j}\int_{a(1+\delta\ell)>|r_1-r_2|>a} \langle\phi_k((r_1+r_2)/2)\rangle\frac{d^d r_1 d^d r_2}{a^{2d-x_i-x_j}}.$$

$$(5.11)$$

The integral gives a factor $S_d a^d \delta\ell$, where S_d is the area of a hypersphere of unit radius in d dimensions.† This term may then be compensated by making the change

$$g_k \to g_k - \tfrac{1}{2} S_d \sum_{ij} c_{ijk} g_i g_j \, \delta\ell. \qquad (5.12)$$

In fact, because of the structure of the operator product expansion (the fact that it is independent of all the other operators Φ in (5.3)), a similar change works to all orders in the expansion (5.8). It is simply a matter of keeping track of the combinatorics to verify that this is the case. Finally, there is the dependence on a which arises through the finite size L. However, it is important *not* to change L, otherwise we should have done nothing at all! As a result, $N^{-1} \equiv a/L$ flows according to the finite-size scaling arguments of Section 4.4.

Putting together the two contributions to the renormalization of the coupling constants,

$$dg_k/d\ell = (d - x_k) g_k - \tfrac{1}{2} S_d \sum_{ij} c_{ijk} g_i g_j + \cdots. \qquad (5.13)$$

The renormalization group eigenvalues at the trivial fixed point are $y_k = d - x_k$, in accord with the fundamental scaling relation (3.55). The somewhat unpleasant factor of $\tfrac{1}{2} S_d$ may be removed by a simple uniform rescaling $g_k \to (2/S_d) g_k$ of the couplings, and this we shall assume to have already been done from now on. The final form of the perturbative renormalization group equations is therefore

$$\boxed{\; dg_k/d\ell = y_k g_k - \sum_{ij} c_{ijk} g_i g_j + \cdots \;} \qquad (5.14)$$

Note that we have not assumed that the scaling operators are normalised in any particular manner in deriving this (merely that this normalisation is independent of a). In practical applications, it is useful to have this freedom. The higher order terms on the right hand side arise from situations in which three or more integration points approach within a distance a of each other. Such contributions are evidently very difficult to evaluate using the hard core

---

† Explicitly $S_d = 2\pi^{d/2}/\Gamma(d/2)$, although its value will never be needed in the sequel.

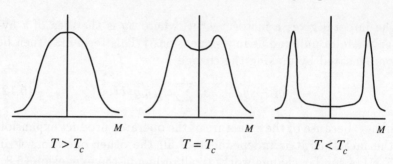

Figure 5.3. Schematic histograms of the block magnetisation in high dimensions.

cut-off used above. At this point, more sophisticated approaches, such as the use of renormalized quantum field theory, show their advantage. However, a considerable amount of formal apparatus needs to be erected before these methods yield useful results, and this goes beyond the remit of this volume.

5.3 The Ising model near four dimensions

We shall now apply the above methods to the well-known and important example of the critical Ising model near four dimensions. In higher dimensions, block spin transformations of the type described in Section 3.1 become less and less reliable. One reason for this (apart from the proliferation of possible couplings) is seen if one studies typical histograms of the total magnetisation $\sum_i s_i$ of a block immersed in a large system, shown schematically in Figure 5.3. Close to the critical point, the distribution gets less and less peaked around two values as the dimension increases. It therefore becomes less reasonable to model this situation with block spins which take only the values ± 1. Instead, it is more appropriate to consider *continuous spins* $S(r)$, which are real variables which can, in principle, take any value. However, we expect their distribution still to be peaked around the typical values ± 1. One way to model this is to write down a continuous spin hamiltonian, of the form

$$\mathcal{H} = -\tfrac{1}{2}\sum_{r,r'} J(r - r')S(r)S(r') - H\sum_r S(r) + \lambda\sum_r (S(r)^2 - 1)^2,$$

$$(5.15)$$

where λ is a number of order unity. The partition function is now

$$Z = \int \prod_r dS(r)\, e^{-\mathcal{H}}.$$

$$(5.16)$$

This modification of the original problem may seem quite barbaric and arbitrary, but, in the spirit of universality, it is quite justified as long as we remain within the domain of attraction of the same critical fixed point. In the end, the justification for this assumption has to be in a comparison of the final answers with those obtained by other methods. In addition, we must be content with calculating only universal quantities, and not, for example, the critical temperature. Since the microscopic parameters for real systems are, in any case, not completely known, in practice this is no great limitation.

An advantage of using a continuous spin variable is that the space continuum limit is more transparent. Since we are interested in only the long wavelength modes of $S(r)$, we may further approximate

$$\sum_{r,r'} J(r - r')S(r)S(r') \approx$$

$$\sum_{r,r'} J(r - r')S(r)\left[S(r) + (r - r')\cdot\nabla S + \tfrac{1}{2}(r - r')^2\nabla^2 S + \cdots\right]$$

$$= J\sum_r \left(S(r)^2 - R^2 a^2(\nabla S(r))^2\right) + \cdots,$$

$$(5.17)$$

after symmetrisation and integration by parts. As in Chapter 2, $J = \sum_r J(r)$ and $R^2 J = \sum_r r^2 J(r)$. The last step is to approximate the sum on r by an integral. The resultant hamiltonian has the form

$$\mathcal{H} = \int \left[\tfrac{1}{2}Ja^2 R^2(\nabla S)^2 - (2\lambda + J)S^2 + \lambda S^4 - HS\right]\frac{d^d r}{a^d}.$$

$$(5.18)$$

Finally, since S is a continuous variable, we may rescale it so that the coefficient of the $(\nabla S)^2$ term is $\tfrac{1}{2}$: $S(r)^2 \to (a^{d-2}/JR^2)S(r)^2$.

The hamiltonian may then be written as

$$\mathcal{H} = \int [\tfrac{1}{2}(\nabla S)^2 + ta^{-2}S^2 + ua^{d-4}S^4 + ha^{-d/2-1}S]d^d r, \quad (5.19)$$

where† $t = -(2\lambda + J)/JR^2$, $u = \lambda/J^2R^4$, and $h = -H/J^{1/2}R$. The form of the hamiltonian in (5.19) is sometimes called the Landau–Ginzburg–Wilson model.

There are three dimensionless couplings here: t, h and u. We may trade J, R and H for them as coordinates in the space of all couplings. The infinite set of other couplings may now be thought of as multiplying all the other terms involving higher powers of S and higher numbers of derivatives, neglected in writing (5.19). As will be seen, these are all irrelevant near four dimensions. We may now apply the renormalization group to lowest order by letting $a \to ba$, and asking how we should change the parameters so as to keep (5.19) invariant. This is simply

$$t' = b^2 t \tag{5.20}$$

$$h' = b^{d/2+1}h \tag{5.21}$$

$$u' = b^{4-d}u. \tag{5.22}$$

5.4 The Gaussian fixed point

There is an obvious fixed point of these equations at $t = h = u = 0$. This is called the *Gaussian* fixed point, because the fixed point hamiltonian is then $\mathcal{H}^* = \tfrac{1}{2}\int(\nabla S)^2 d^d r$, and the partition function is simply a Gaussian integral. There is an obvious difference in the properties of this fixed point depending on whether d is greater than, or less than, four.

For $d > 4$, u is irrelevant, and the fixed point has two relevant variables, one in the thermal sector and one magnetic, as expected of a fixed point which is to describe the critical Ising model. The eigenvalues are simply $y_t = 2$ and $y_h = d/2 + 1$, and, following the arguments in Section 3.5 we may then deduce the values of the various exponents. These are shown in Table 5.1, together with predictions of mean field theory, which, as we argued in Section 2.4, should give the correct exponents for $d > 4$. Note that

---

† The reader who is puzzled why the value of t is apparently always negative is referred to the discussion on p.95.

Table 5.1. *Exponents at the Gaussian fixed point and in mean field theory.*

Exponent	Gaussian	MFT
α	$2 - d/2$	0
β	$(d-2)/4$	$\frac{1}{2}$
γ	1	1
δ	$(d+2)/(d-2)$	3
ν	$\frac{1}{2}$	$\frac{1}{2}$
η	0	0

agreement is perfect when $d = 4$, and that the exponents (γ, ν, η) which may be related to the correlation function all agree.‡ But something is wrong in the free energy. For $d > 4$, the predictions of mean field theory are all more singular than those of the Gaussian fixed point. In addition, the hyperscaling relation (3.52) is seen to fail.

This failure of conventional renormalization group scaling may be understood in several ways. In terms of the homogeneous scaling law (3.40) for the singular part of the free energy, it must be that u is *dangerously* irrelevant, that is, it may not simply be set equal to zero (see p.49). This is most spectacularly seen in the spontaneous magnetisation: if we simply minimise the hamiltonian in (5.19) we find that $\langle S \rangle \propto (-t/u)^{1/2}$. This is consistent with the mean field result $\beta = \frac{1}{2}$, but it certainly makes no sense to set $u = 0$ in this result.

Alternatively, we can consider the free energy as calculated by the iterated inhomogeneous transformation law (3.60). The sum consists of two parts, one coming from the vicinity of the Gaussian fixed point, which will give rise to contributions scaling with the exponents in the second column of Table 5.1, and a piece from the renormalization group trajectory far from the fixed point, where mean field theory should apply. Since the mean field values are more singular than the Gaussian ones, this contribution actually dominates. This argument does, however, suggest that the Gaus-

‡ Although the susceptibility is given by a suitable derivative of the free energy, it is also proportional to $\sum_r \langle s(r)s(0) \rangle$.

sian values should be observed as subleading corrections to the mean field terms – a feature which is not present in unadorned mean field theory.

When $d < 4$, u is relevant at the Gaussian fixed point, and this is no longer the appropriate one to describe the critical point of the Ising model. However, if the value of u happens to be small in a particular system, we may use the cross-over theory of Section 4.2. The cross-over exponent is $\phi = (4-d)/2$, and the cross-over region is given by $|t| \sim u^{2/(4-d)}$. If we express this in terms of ξ and R we then recover the Ginzburg criterion of Section 2.4.

5.5 The Wilson–Fisher fixed point

When the dimension d is slightly smaller than four, the scaling field u is only slightly relevant at the Gaussian fixed point, and therefore we might hope to find another nearby to which it flows, using the perturbative methods described in Section 5.2. In order to do this it is necessary to compute the operator product expansion coefficients at the Gaussian fixed point. We first have to learn how to compute correlation functions in the Gaussian model, with hamiltonian $\mathcal{H} = \frac{1}{2} \int (\nabla S)^2 d^d r$. Fortunately this is very simple. Since the reduced hamiltonian must be dimensionless, S must have dimensions $(\text{length})^{-d/2+1}$. Therefore the 2-point correlation function $\langle S(r_1)S(r_2) \rangle$ must be proportional to $|r_1 - r_2|^{-d+2}$. Since we may always redefine the normalisation of S so that the coefficient of this power law is unity, we therefore take

$$\langle S(r_1)S(r_2) \rangle = \frac{1}{|r_1 - r_2|^{d-2}}. \tag{5.23}$$

Higher correlation functions $\langle S(r_1) \ldots S(r_{2n}) \rangle$ in the Gaussian model are given by the following rule (see Appendix): sum over all possible ways of dividing the set of points $(r_1, \ldots, r_{2n})$ into n pairs, each pair corresponding to a two-point correlation function, or *Wick contraction*. For example

$$\langle S(r_1)S(r_2)S(r_3)S(r_4) \rangle = \langle S(r_1)S(r_2) \rangle \langle S(r_3)S(r_4) \rangle + $$
$$\langle S(r_1)S(r_3) \rangle \langle S(r_2)S(r_4) \rangle + \langle S(r_1)S(r_4) \rangle \langle S(r_2)S(r_3) \rangle. \tag{5.24}$$

If we decide to represent the two-point function $\langle S(r_1)S(r_2) \rangle$ by a line connecting r_1 and r_2, the terms on the right hand side of (5.24) are represented by the diagrams in Figure 5.4. The above

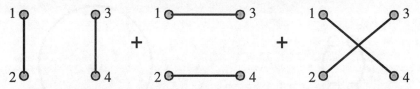

Figure 5.4. Diagrams for the 4-point correlation function in the Gaussian model.

Figure 5.5. Diagrams for the $\langle S^2 S^2 \rangle$ correlation function.

rule is equivalent to *Wick's theorem* in quantum field theory. It still holds when some of the points coincide. To enumerate all the possible pairings in this case, it is convenient to imagine the points as slightly separated, and then take the limit of coincidence at the end. For example, to compute $\langle S^2(r_1)S^2(r_2)\rangle$, we split the points as shown in Figure 5.5. There are three distinct ways of contracting these points in pairs. The first two each give rise to a contribution $\langle S(r_1)S(r_2)\rangle^2$. The third is different, however, and gives $\langle S^2 \rangle^2$, independent of $|r_1 - r_2|$. Such terms do not contribute to the renormalization group equations, and it is convenient to remove them by defining *normal ordered* operators by

$$:S^2:\equiv S^2 - \langle S^2 \rangle \tag{5.25}$$

$$:S^4:\equiv S^4 - 3\langle S^2 \rangle S^2, \tag{5.26}$$

and so on. This merely corresponds to taking a slightly different basis of scaling operators at the Gaussian fixed point. In particular, the coefficient of $:S^2:$ in the hamiltonian changes according to $t \to t' = t + 3ua^{d-2}\langle S^2 \rangle$, which now changes sign as the temperature is varied. With the normal ordered operators, Wick's theorem is now modified by the rule: ignore all diagrams in which both ends of a line are connected to the same point. This considerably reduces the number of diagrams, without losing any of the physics.

Wick's theorem also applies to the operator product expansion. For example, consider the operator product $:S(r_1)^2 :: S(r_2)^2 :$. This means that we should consider all possible correlation functions

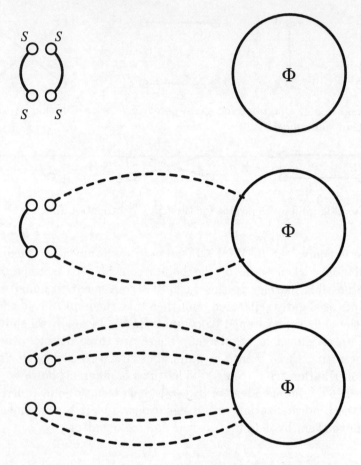

Figure 5.6. Operator product expansion of S^2 and S^2.

of the form $\langle : S(r_1)^2 :: S(r_2)^2 : \Phi \rangle$. Applying Wick's theorem to this gives a number of terms, depending on how many of the $S(r_1)$ are connected to the $S(r_2)$. These possibilities are illustrated in Figure 5.6. If both the operators at r_1 are connected to operators at r_2, there is no further possibility of connection to the operators in Φ. Thus, from the point of view of Φ, this pairing will behave as the insertion of the unit operator at r_1. Similarly, if just one of the $S(r_1)$ is connected onto one of the $S(r_2)$, this leaves a product $S(r_1)S(r_2)$ each of whose factors must be connected onto an operator in Φ. From a distance, this operator will then look like

$:S(r_1)^2:$. Finally, if none of the operators at r_1 or r_2 are connected to each other, they must all be connected to operators in Φ, and, from a distance, they will look like $:S(r_1)^4:$. Once again, using normal ordering allows us to neglect all contributions where operators at the same point are connected. Proceeding in this manner, we find that the required operator product expansion has the form

$$:S(r_1)^2::S(r_2)^2:=\frac{2}{r_{12}^{2d-4}}+\frac{4}{r_{12}^{d-2}}:S(r_1)^2:+:S(r_1)^4:+\cdots,$$

(5.27)

where $r_{12}=|r_1-r_2|$, and the neglected terms include derivative operators. The only important quantities in (5.27) are the operator product expansion coefficients $(2,4,1,\ldots)$, and we see that they are very simply determined by counting arguments. It is similarly straightforward to evaluate the other operator product expansions. Denoting $:S^n:$ by ϕ_n, and focussing only on the coefficients themselves, the relevant operator product expansions may be written schematically as

$$\phi_1\cdot\phi_1=1+\phi_2 \tag{5.28}$$

$$\phi_1\cdot\phi_2=2\phi_1+\phi_3 \tag{5.29}$$

$$\phi_1\cdot\phi_4=4\phi_3+\cdots \tag{5.30}$$

$$\phi_2\cdot\phi_2=2+4\phi_2+\phi_4 \tag{5.31}$$

$$\phi_2\cdot\phi_4=12\phi_2+8\phi_4+\cdots \tag{5.32}$$

$$\phi_4\cdot\phi_4=24+96\phi_2+72\phi_4+\cdots. \tag{5.33}$$

The neglected terms in the above correspond to values of $n>4$, and to derivative operators, both of which will turn out to be irrelevant. Notice that the operator $\phi_3=:S^3:$ has been generated. For reasons to be explained below on p.101, this may be ignored. In fact, very few of the operator product expansion coefficients shown in (5.28–5.33) actually enter into the subsequent calculation. To illustrate further the computation of one which does, consider the coefficient of ϕ_4 in the operator product $\phi_4\cdot\phi_4$. One of the contractions is shown in Figure 5.7. Two of the factors $S(r_1)$ must be connected onto two of the $S(r_2)$, leaving four S's over. The two $S(r_1)$ may be chosen in $4\cdot3$ ways, and similarly for the two $S(r_2)$. But this method of reasoning will double count, since there are $2!$ equivalent permutations of the lines which connect r_1 and r_2. Thus the correct coefficient is $(4\cdot3)^2/2=72$.

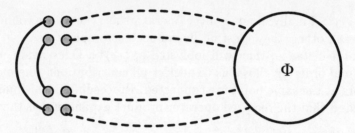

Figure 5.7. Contribution to S^4 in the operator product expansion of S^4 with S^4.

We are now in a position to write down the renormalization group equations, using (5.14):

$$dh/d\ell = (d/2 + 1)h - 2 \cdot 2ht + \cdots \tag{5.34}$$

$$dt/d\ell = 2t - h^2 - 4t^2 - 2 \cdot 12tu - 96u^2 + \cdots \tag{5.35}$$

$$du/d\ell = \epsilon u - t^2 - 2 \cdot 8tu - 72u^2 + \cdots, \tag{5.36}$$

where $\epsilon = 4 - d$. The method for obtaining these is straightforward: for example, to find the terms contributing to $dt/d\ell$ we simply look at where ϕ_2 occurs on the right hand side of the operator product expansions in (5.28–5.33). Note that contributions where $i \neq j$ in (5.14) are counted twice, since $c_{ijk} = c_{jik}$, and therefore appear with factors of two.

If we now suppose that ϵ is small, we see that, apart from the Gaussian fixed point, there is another in which $h = 0$, $u = u^* = \epsilon/72 + O(\epsilon^2)$, and $t = O(\epsilon^2)$. This is called the *Wilson–Fisher fixed point*. The thermal eigenvalue at this new fixed point is given by linearising (5.35):

$$dt/d\ell = 2t - 24u^*t + \cdots, \tag{5.37}$$

so that $y_t = 2 - (24/72)\epsilon + O(\epsilon^2)$. Finally, this gives for the correlation length exponent $\nu = 1/y_t$

$$\boxed{\nu = \tfrac{1}{2} + \tfrac{1}{12}\epsilon + O(\epsilon^2)} \tag{5.38}$$

We see that the $O(\epsilon)$ corrections to the exponents are given by simple ratios of operator product expansion coefficents at the Gaussian fixed point. In fact, as stated earlier, very few of the coefficients calculated in (5.28–5.33)) enter into the final result. With hindsight, we could therefore have saved even more effort. Because ϕ_1 does not appear in the operator product expansion of ϕ_1

Table 5.2. *Estimates for critical exponents of the two-dimensional Ising model from the resummed ϵ-expansion at order ϵ^5, compared with exact results.*

	ϵ-expansion	Exact
ν	0.99 ± 0.04	1
η	0.26 ± 0.05	0.25

and ϕ_4, the $O(\epsilon)$ correction to the magnetic exponent vanishes, so $\eta = 0$ to this order. In fact, it does receive a contribution at $O(\epsilon^2)$.

The usefulness of the ϵ-expansion

The computation of the critical exponents beyond the first order requires knowledge of the higher order terms in the renormalization group equations (5.14). These do not follow simply from the operator product expansion, and require more sophisticated calculations of Feynman diagrams which are beyond the scope of this book. However, in any case, one might question the numerical accuracy of an expansion scheme in powers of a parameter ϵ which ultimately has to be set equal to one (or two) to obtain results of direct physical significance. In fact, in many cases, the ϵ-expansion does remarkably well if carried out to sufficiently high orders. It is believed, however, that it is only an asymptotic expansion. That is, for any non-zero value of ϵ, no matter how small, the series will eventually give inaccurate results if truncated at too high an order. There are, nevertheless, a number of efficient resummation methods available which yield strongly convergent estimates, even for $d = 2$. The best values for the Ising model ($n = 1$) universality class in two and three dimensions are shown in Tables 5.2 and 5.3, and compared with the results of other accurate methods: a direct field-theoretic renormalization group calculation, high-temperature expansions, and exact results.†

Even in other cases where the ϵ-expansion fares poorly from a

---

† There appears to be, in fact, a small but persistent discrepancy between the high-temperature expansion estimates in three dimensions and those of the RG, which has not been satisfactorily explained.

Table 5.3. *Estimates for critical exponents of the three-dimensional Ising model from the resummed ϵ-expansion at order ϵ^5, compared with resummed direct RG and high-temperature expansion methods. The last estimate for η follows from those for γ and ν and the scaling relation*
$$\gamma = \nu(2 - \eta).$$

	ϵ-expansion	Direct RG	High-T expansion
ν	0.6305 ± 0.0025	0.6300 ± 0.0015	0.633 ± 0.0013
η	0.037 ± 0.003	0.031 ± 0.004	0.042 ± 0.005

numerical point of view, it nevertheless provides an unambiguous classification of possible fixed points in a given system, the number and nature of the various relevant scaling variables at each fixed point, and the existence of scaling laws between the various critical exponents. All of these are robust predictions of the method which hold independently of its ability to make quantitative statements. This should be contrasted with real space methods where such results may depend strongly on the precise transformation chosen.

Irrelevant operators

We now discuss an important point which has been ignored so far. The appearance of the S^4 term in the continuous spin hamiltonian was crucial to understanding the role of four dimensions, yet it seemed to emerge from the somewhat arbitrary choice of $(S^2 - 1)^2$ as a weight function in (5.15). What if we had chosen some other? In general, such a weight function should be an even function of S, and we may therefore consider expanding it as a power series $\sum_n g_{2n} S^{2n}$. If we make the same rescaling of S as before, in the continuum version the coupling g_{2n} will come along with a factor $a^{n(d-2)-d}$. Thus, at the Gaussian fixed point where all the g_{2n} vanish, this coupling has a renormalization group eigenvalue $2n - (n-1)d$. If d is near 4, then all couplings with $2n > 4$ are irrelevant at the Gaussian fixed point, so that they give only correction to scaling terms. Since the non-trivial fixed point is, for small ϵ, close to the Gaussian fixed point, we expect the eigenvalues of these operators to be modified only to $O(\epsilon)$ at worst, and so they will remain irrelevant at the new fixed point.

One might object that g_6 becomes marginal exactly in three dimensions, where we hope eventually to apply the theory, and therefore it may not be neglected. However, this criticism is not well-founded, because, as we showed above, the relevant fixed point for $d < 4$ is not the Gaussian, but rather the Wilson–Fisher fixed point, and we should really ask whether g_6 is relevant there. In order to study this question in greater depth, let us calculate the renormalization group eigenvalue of g_6 to first order in ϵ. According to (5.14) this involves evaluating the coefficient of ϕ_6 appearing in the product of ϕ_4 and ϕ_6. By the same counting arguments as above, this is $4!6!/4!2!2! = 180$, so the renormalization group equation is

$$dg_6/d\ell = (6 - 2d)g_6 - 360ug_6 + \cdots, \qquad (5.39)$$

where none of the neglected terms affect the $O(\epsilon)$ calculation. Expanding in ϵ we then see that the eigenvalue at the non-trivial fixed point is

$$-2 + 2\epsilon - 5\epsilon + O(\epsilon^2), \qquad (5.40)$$

where the third term comes from the $O(u^*)$ correction. We see that this quite overwhelms the second term, and that u_6 is actually *less* relevant in $4 - \epsilon$ dimensions than in $d = 4$! Of course, such a result might not survive to higher orders in ϵ, but all indications are that it does.

One consequence of this calculation is that the corrections to scaling terms coming from the higher powers of S are not expected to be very important. The dominant correction to scaling comes from the irrelevant variable $u^* - u$, which has eigenvalue $y = -\epsilon + O(\epsilon^2)$. This will give rise to corrections to scaling of the order $|t|^{|y|/y_t} \approx |t|^{1/2}$ in three dimensions.

Redundant operators

Let us now return to the question of the S^3 term, which according to (5.30), is generated by the renormalization group when $h \neq 0$, even if it is not present in the original hamiltonian (5.19). Power counting would suggest that such an operator is relevant near $d = 4$; how, then, may it be ignored? If we explicitly include such a term in the hamiltonian, it then has the form

$$\mathcal{H} = \int [\tfrac{1}{2}(\nabla S)^2 + tS^2 + u_3 S^3 + uS^4 + hS]d^d r, \qquad (5.41)$$

where the factors of a have been suppressed for clarity. It is now clear that the S^3 term may be removed by a suitable shift $S \to S + \text{const.}$, at the cost of redefining t and h. Since S is nothing but an integration variable over which the trace is performed in the partition function, we are clearly at liberty to make such a shift without altering the physics in any way. In fact, if we make an infinitesimal uniform shift $S \to S + \delta S$, the first order change in the hamiltonian density is

$$(2tS + 3u_3 S^2 + 4uS^3 + h)\delta S, \qquad (5.42)$$

which corresponds to a redefinition of the parameters

$$h \to h + 2t\delta S \qquad (5.43)$$
$$t \to t + 3u_3 \delta S \qquad (5.44)$$
$$u_3 \to u_3 + 4u\delta S. \qquad (5.45)$$

Thus, within the 4-dimensional subspace of renormalization group flows parametrised by (h, t, u_3, u), there is a family of one-dimensional curves generated by (5.43–5.45) along which the actual physics does not change. The tangent vectors to these curves are called redundant scaling variables, and the operators to which they couple in the hamiltonian (in this case $2tS + 3u_3 S^2 + 4uS^3$) are called *redundant operators*.

Thus, although S^3 is apparently relevant, it may be eliminated by a suitable change in h and t. Since this shift is itself of $O(u)$, however, it does not affect the calculation of the critical exponents to first order in ϵ. The redundant operator described above is not the only one. In fact, any suitably local redefinition of the variable S will correspond to such a direction in parameter space. Within the even subspace, such a redefinition has the form

$$S \to S + a_1 S^3 + a_2 S^5 + \cdots + b_1 \nabla^2 S + b_2 S^2 \nabla^2 S + \cdots \qquad (5.46)$$

and it may be seen to introduce only operators which are irrelevant near $d = 4$. Because of this redundancy, the physical space of irrelevant operators is smaller than it might appear, but, it is still very large! In the Ising universality class, only the S^3 operator is superficially relevant but also redundant.

5.6 Logarithmic corrections in $d = 4$

When $d = 4$, the scaling variable u is marginal. While $d = 4$ is not physically relevant, it does provide a simple example in which the consequences of the existence of such a marginal variable may be illustrated. For $d = 4$, the renormalization group equations reduce to

$$du/d\ell = -72u^2 + \cdots \qquad (5.47)$$

$$dt/d\ell = 2t - 24ut + \cdots, \qquad (5.48)$$

where all the inessential terms have been dropped. We see that u is in fact marginally *irrelevant*, so that the large distance behaviour is still controlled by the Gaussian fixed point. However, as u will flow to zero only very slowly, we might expect modifications of the behaviour inferred from the Gaussian fixed point. This is indeed the case.

The equation for u may be integrated straightforwardly to give

$$u(\ell) = \frac{u(0)}{1 + 72u(0)\ell}. \qquad (5.49)$$

The transformation law (3.26) for the singular part of the free energy, suitably generalised to the case of an infinitesimal renormalization group transformation, implies that

$$f_s(t, u) = e^{-d\ell} f_s\left(t(\ell), u(\ell)\right). \qquad (5.50)$$

As usual, we choose $\ell = \ell_0$, so that $t(\ell_0) = t_0 = O(1)$. We may find ℓ_0 by observing that

$$\ln(t_0/t) = \int_t^{t_0} \frac{dt'}{t'} = \int_0^{\ell_0} \left[2 - \frac{24u(0)}{1 + 72u(0)\ell}\right] d\ell$$

$$= 2\ell_0 - \tfrac{1}{3}\ln\left[1 + 72u(0)\ell_0\right]. \qquad (5.51)$$

This transcendental equation cannot be solved explicitly, but, for large t_0/t, it may be iterated to give the asymptotic expansion

$$\ell_0 = \tfrac{1}{2}\ln(t_0/t) + \tfrac{1}{6}\ln\left[1 + 36u(0)\ln(t_0/t)\right] + \cdots. \qquad (5.52)$$

From (5.50)

$$f_s(t, u(0)) = e^{-4\ell_0} f_s(t_0, u(\ell_0)). \qquad (5.53)$$

In estimating the right hand side we should be mindful that u may be a dangerous irrelevant variable, in which case it is not permissible simply to set $u(\ell_0) = 0$. Since the right hand side is to be evaluated outside the critical region, we may use the mean field results of Chapter 2 to estimate that $f_s(t_0, u(\ell_0)) \propto t_0^2/u(\ell_0)$. Using the value of $u(\ell_0)$ from (5.49) then yields

$$f_s(t, u(0)) \propto t^2 \left[1 + 36u(0)\ln(t_0/t)\right]^{1/3}, \qquad (5.54)$$

When $u(0) = 0$, we recover the results of the Gaussian fixed point, but for $u(0) \neq 0$ there are multiplicative *logarithmic corrections* to scaling. In particular, for the specific heat we find

$$C \propto |\ln(t/t_0)|^{1/3}, \qquad (5.55)$$

as $t \to 0$. This type of behaviour is a typical consequence of the existence of a marginally irrelevant variable. Note that the power of the logarithm in (5.55) was simply determined by the ratio of operator product expansion coefficients.

Although the above example was in four dimensions, there are physical examples of universality classes which have $d_c = 3$ as their upper critical dimension, and which are therefore expected to exhibit such logarithmic corrections. Examples are the tricritical Ising model (see Ex. 2.6), the theta-point in linear polymers (see Section 9.5), and uniaxial dipolar ferromagnets. However, such logarithmic corrections are very difficult to see in practice, either experimentally or in numerical simulations. The asymptotic behaviour like that in (5.55) is apparent only when $36u(0)\ln(t_0/t) \gg 1$, and, since $u(0)$ is not universal, it is difficult to predict when this will set in. Even trying to fit the data with the more accurate form (5.54) can be misleading, because the higher terms in the asymptotic expansion (5.52) lead to contributions which are down only by powers of $\ln(t_0/t)$, rather than powers of t as in the normal case.

5.7 The $O(n)$ model near four dimensions

So far we have considered only the universality class of the critical Ising model. All the above calculations may be generalised quite easily to the $O(n)$ model or n-vector model introduced in Section 1.2. Instead of Ising spins taking the values ± 1, these models have n-component spins $\mathbf{s}(r)$, normalised so that $\mathbf{s}(r)^2 = 1$. The continuous spin version of this is a simple generalisation of that for the Ising model ($n = 1$):

$$\mathcal{H} = \int \left[\tfrac{1}{2}(\nabla \mathbf{S})^2 + ta^{-2}\mathbf{S}^2 + ua^{d-4}(\mathbf{S}^2)^2 + ha^{-d/2-1}S_1\right]d^d r. \qquad (5.56)$$

The ϵ-expansion for this model is very similar to that carried out in the Ising case. The only difference lies in the values of the various operator product expansion coefficients.

In the Gaussian model, the various components of **S** are decoupled, so the 2-point correlation function has the form

$$\langle S_i(\vec{r}_1) S_j(\vec{r}_2) \rangle = \frac{\delta_{ij}}{r_{12}^{d-2}}. \tag{5.57}$$

As in the Ising case, there are only two important operator product expansion coefficients which enter the calculation to first order in ϵ. The first is the coefficient of $S_i S_i$ in the product of $S_i S_i$ and $(S_j S_j)(S_k S_k)$ (where the summation convention is assumed). There are two types of contribution, one where both S_i are connected either to both S_j or to both S_k, and another where one S_i is connected to an S_j and one to an S_k. The first type of contribution involves $\delta_{ij} \delta_{ij} = n$, and it is not difficult to see that the coefficient is 4. The second is independent of n. Its evaluation is rather more complicated, but in fact we have to do no more work, because we already know from the previous section that the result is 12 when $n = 1$. For arbitrary n, then, this operator product expansion coefficient must be $4(n + 2)$. A similar method works for the coefficient of $(\mathbf{S}^2)^2$ in its operator product with itself. In this case, the term proportional to n is easily seen to be $8n$, and so, comparing with the case $n = 1$, we find the general result $8(n+8)$.

The important terms in the renormalization group equations for the $O(n)$ model are therefore

$$du/d\ell = \epsilon u - 8(n + 8)u^2 \cdots \tag{5.58}$$
$$dt/d\ell = 2t - 8(n + 2)ut + \cdots. \tag{5.59}$$

The fixed point is now at $u^* = \epsilon/8(n+8)$, and the thermal eigenvalue at the non-trivial fixed point is

$$y_t = \nu^{-1} = 2 - \frac{n+2}{n+8}\epsilon + O(\epsilon^2). \tag{5.60}$$

The fact that $u^* \to 0$ as $n \to \infty$ suggests that the above result may be exact in this limit. In fact, this turns out to be the case, as may be verified by an independent calculation as follows. For large n, $\mathbf{S}^2 = \sum_a S_a^2$ is the sum of a large number of random variables and, by the central limit theorem, it should have a normal distribution. In particular, we may replace the interaction term $(\mathbf{S}^2)^2$ in

(5.56) by its cumulant $3\langle \mathbf{S}^2 \rangle \mathbf{S}^2$. This reduces the hamiltonian to a Gaussian form, so that $\langle \mathbf{S}^2 \rangle$ may be calculated self-consistently using the rules given in the Appendix:

$$\langle \mathbf{S}^2 \rangle = n \int_{\text{BZ}} \frac{1}{k^2 + t_0 + 3u\langle \mathbf{S}^2 \rangle} \frac{d^d k}{(2\pi)^d}. \tag{5.61}$$

A subscript has been added to the reduced temperature to stress that it is defined relative to the *mean field* critical temperature. This equation in principle determines $\langle \mathbf{S}^2 \rangle$ for each value of t_0. Once this is solved, the correlation length is then given by $\xi^{-2} = t + 3u\langle \mathbf{S}^2 \rangle$. Therefore (5.61) may be rewritten as

$$\xi^{-2} = t_0 + 3nu \int_{\text{BZ}} \frac{1}{k^2 + \xi^{-2}} \frac{d^d k}{(2\pi)^d} \tag{5.62}$$

$$= t_0 + 3nu \int_{\text{BZ}} \frac{1}{k^2} \frac{d^d k}{(2\pi)^d} - 3nu \int_{\text{BZ}} \frac{\xi^{-2}}{k^2(k^2 + \xi^{-2})} \frac{d^d k}{(2\pi)^d}.$$

Setting $\xi^{-2} = 0$ determines the actual critical temperature. Note that the fluctuations depress it below the mean field value, as expected. When $d > 4$, the last integral is finite when $\xi^{-2} = 0$, so that

$$\xi^{-2} = t_0 - t_{0c} + O(\xi^{-2}). \tag{5.63}$$

As a result $\xi^{-2} \propto t = t_0 - t_{0c}$, corresponding to the expected mean field value $\nu = \frac{1}{2}$. However, when $d < 4$ this is no longer permissible. Instead, we may argue that the ultraviolet cut off on large k is no longer relevant, and it may be removed. In that case, simple power counting shows that, close to the critical point,

$$\xi^{-2} = t - \text{const.} \, \xi^{2-d} + \cdots, \tag{5.64}$$

at least when $2 < d < 4$. In that case, the left hand side is asymptotically negligible, and

$$\nu = 1/(d-2), \tag{5.65}$$

in agreement with (5.60). Note that $\nu \to \infty$ as $d \to 2+$. This is related to the fact that $d = 2$ is the lower critical dimensionality for this model (in fact, for all $n > 2$), as we shall see in Section 6.1. Although it has no experimental relevance, the $n \to \infty$ limit provides a non-trivial testing ground for theoretical approaches, and forms the starting point for the $1/n$ expansion, which gives an approximation scheme complementary to the ϵ-expansion.

5.8 Cubic symmetry breaking

As a final introductory example of the power of the perturbative renormalization group, we consider the case of cubic symmetry breaking fields acting on the $O(n)$ model above. This is a famous example of the power of the renormalization group analysis in a situation where it predicts rather different behaviour from that suggested by mean field theory. Cubic fields are important for the discussion of, for example, XY or Heisenberg magnets where the magnetic ions are arranged in a crystal of cubic symmetry. In that case, crystalline fields will break the symmetry of the hamiltonian down from the full $O(n)$ rotational group acting on the spin vector $\mathbf{S}$, to its cubic subgroup. This allows further terms to appear in the continuous spin hamiltonian, consistent with the reduced symmetry. In principle, these could be very complicated. However, near four dimensions, we need consider only terms up to and including fourth order in S, since higher powers would be irrelevant. In addition to the terms already present in the $O(n)$ model, the only other allowed term has the form $v \sum_i S_i^4$, so that the relevant hamiltonian is (suppressing powers of a)

$$\mathcal{H} = \int \left[\tfrac{1}{2}(\nabla S)^2 + t \sum_i S_i^2 + u \sum_{ij} S_i^2 S_j^2 + v \sum_i S_i^4 \right] d^d r. \quad (5.66)$$

To write the first order renormalization group equations for u and v, we need the appropriate operator product expansion coefficients. The operator product of $\sum_{ij} S_i^2 S_j^2$ with itself has already been computed in the previous section: it is $8(n+8)$. The term $\sum_i S_i^4$ behaves, at the Gaussian fixed point, like a sum of n decoupled Ising-like S^4 terms. Therefore its operator product expansion with itself can only give this term back, with the same coefficient of 72 we found in the Ising case. The only new operator product expansion which need be evaluated, therefore, is that of $\sum_{ij} S_i^2 S_j^2$ with $\sum_k S_k^4$. There are two types of contribution. The first is where either both S_i or both S_j are connected to an S_k: there are $2 \cdot 2 \cdot 4 \cdot 3/2 = 24$ ways of doing this, and the residual operator has the form $\sum_{ijk} \delta_{ik}\delta_{ik} S_j^2 S_k^2$, so this contributes to the renormalization of u. In the second type, one S_i and one S_j each connect to an S_k: there are $2 \cdot 2 \cdot 4 \cdot 3 = 48$ ways of doing this. The remaining factors have the form $\sum_{ijk} \delta_{ik}\delta_{jk} S_i S_j S_k^2$, so this contributes to the renormalization of v. The renormalization group equations for the couplings u and v are therefore

$$du/d\ell = \epsilon u - 8(n+8)u^2 - 48uv + \cdots \qquad (5.67)$$
$$dv/d\ell = \epsilon v - 96uv - 72v^2 + \cdots, \qquad (5.68)$$

where the neglected terms are higher order in u, v and ϵ. Apart from the Gaussian fixed point at $u = v = 0$, these equations have three non-trivial fixed points of $O(\epsilon)$:

- $(u, v) = (\epsilon/8(n+8), 0)$. This is the usual fixed point of the $O(n)$ model, in the absence of cubic fields;
- $(u, v) = (0, \epsilon/72)$ corresponds to n decoupled Ising models;
- $(u, v) = (\epsilon/24n, (n-4)\epsilon/72n)$. This is new, and is called the *cubic fixed point*.

In order to understand which fixed point controls the critical behaviour, we may perform a linear analysis of their relative stabilities. At the decoupled Ising fixed point, the eigenvalue corresponding to u is $\epsilon - 48v^* = \epsilon/3$, so this fixed point is always unstable. On the other hand, at the $O(n)$ fixed point, the eigenvalue corresponding to v is $\epsilon - 96u^* = (n-4)\epsilon/(n+8)$, which is unstable only when $n > 4$. The scaling variables at the cubic fixed point are non-trivial combinations of $u - u^*$ and $v - v^*$. The eigenvalues are therefore those of the matrix of derivatives

$$\begin{pmatrix} \epsilon - 16(n+8)u^* - 48v^* & -48u^* \\ -96v^* & \epsilon - 144v^* - 96u^* \end{pmatrix}, \qquad (5.69)$$

which gives $(-\epsilon, (4-n)\epsilon/3n)$.

For $n < n_c = 4 - O(\epsilon)$, therefore, cubic symmetry-breaking fields are irrelevant, and lead only to additional corrections to scaling. For $n > n_c$, however, the result depends on the sign of v. For $v > 0$, the critical renormalization group trajectory flows to the cubic fixed point. This is the case illustrated in Figure 5.8. In order to compute the thermal eigenvalue at this new fixed point, we need to consider the renormalization group equation for t. However, both the required operator product expansion coefficients have been calculated already, in the Ising (5.37) and $O(n)$ (5.59) limits. It follows immediately that the renormalization group equation is

$$dt/d\ell = 2t - 8(n+2)ut - 24vt + \cdots, \qquad (5.70)$$

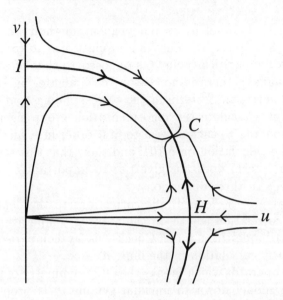

Figure 5.8. RG flows for cubic symmetry-breaking when $n > n_c$.

which gives the thermal eigenvalue as $y_t^{\text{cubic}} = 2 - 2(n-1)\epsilon/3n + O(\epsilon^2)$.

On the other hand, when $v < 0$ initially, the renormalization group trajectories ultimately flow into the region $u + v < 0$, where the renormalized hamiltonian no longer has a stable minimum. In this case, it may be argued that the transition is first order. Such a transition, which would be second order according to mean field theory, is called a *fluctuation-driven* first order transition.

Exercises

5.1 Consider the continuous spin tricritical Ising model, in which the dominant interaction is proportional to $:S^6:$. Show that the upper critical dimension is $d_c = 3$, and calculate the renormalization group eigenvalues to first order in $\epsilon = 3 - d$, using the same method as discussed in Section 5.5. [Note that there are two thermal eigenvalues to be calculated.] How does the specific heat behave in three dimensions?

5.2 Generalise the above calculation to the case of the $O(n)$ model, and verify that the $n \to \infty$ limit of your result agrees with that found by an independent method which replaces the $(S^2)^3$ term by something proportional to $\langle S^2 \rangle^2 S^2$.

5.3   At a fluid critical point (as opposed to that of an Ising ferromagnet), there is no reason to eliminate, on the basis of symmetry, an S^5 term in the corresponding continuous spin model. Calculate the renormalization group eigenvalue corresponding to such a term to first order in ϵ, along the lines of the calculation on p.101, and show that this suggests that such a term does not give rise to important corrections to scaling in three dimensions.

5.4   What are the logarithmic corrections to the dependence of the correlation length on the reduced temperature t for the $O(n)$ model in $d = 4$? Check this by comparison with an explicit calculation in the limit $n \to \infty$.

5.5   The operator which breaks the $O(3)$ symmetry of the Heisenberg model down to uniaxial symmetry is proportional to $S_z^2 - \frac{1}{2}(S_x^2 + S_y^2)$. Calculate its renormalization group eigenvalue, and hence the corresponding cross-over exponent, to first order in ϵ.

5.6   Consider the general case of the cross-over between two nearby fixed points, described to lowest order by the renormalization group equations $du/d\ell = \epsilon u - bu^2$ and, for the external magnetic field, $dh/d\ell = h(y + b'u)$. By solving these equations explicitly and using the transformation law for the correlation function derived on p.51, derive a form for the correlation function which explicitly exhibits the cross-over from a power-law characteristic of one fixed point at short distances, to the other at large distances.

6

Low dimensional systems

6.1 The lower critical dimension

As the dimensionality of space is reduced, the fluctuations around mean field behaviour become more pronounced. These fluctuations act to disorder the system and therefore to lower the critical temperature. Eventually, they may drive it to zero. As we have seen in Chapter 3, for the Ising model (and, more generally, for systems with discrete symmetries) the *lower critical dimension* where this first happens is $d_l = 1$. It turns out, however, that for most systems with continuous symmetry the fluctuations are more severe because the order parameter may easily change its direction with little cost in free energy. As a result, the value of the lower critical dimension is increased to $d_l = 2$.

There exist rather simple arguments based on the free energy cost of destabilising the ordered phase which lead to the above results. Suppose a magnetic field acts on the boundary of the system in such a way as to favour energetically a particular ordered state. What, then, is the cost in the free energy of introducing a domain, of size ℓ, within which the spins are in another possible ground state? Consider first an Ising-like system. In $d = 1$, there will be a finite energy $2J$ associated with each domain wall, but each wall may occupy $O(\ell)$ different positions, so that this set of configurations has an entropy $\sim \ln \ell$. The free energy cost is therefore roughly $4J - 2k_B T \ln \ell$, and so forming such a domain of sufficiently large size will always lower the free energy. Thus the ordered phase cannot be stable. This is a simplified version of Landau's argument for the absence of equilibrium order in one dimension. In $d = 2$, a domain wall of size ℓ will have energy $O(J\ell)$. To estimate its entropy we may imagine it as a closed random walk, which, at each step, on a square lattice, for example, has at most three choices of which way to go, since it must avoid

111

itself. We therefore expect the number of possible configurations to go roughly as μ^ℓ, where $\mu < 3$. As a result the free energy cost is roughly $2J\ell - k_B T \ell \ln \mu$. For sufficiently low temperatures, therefore, the ordered phase should be stable against the formation of large domains of reversed spins. At some temperature $T_c = O(J/k_B)$, this is no longer true and the system breaks up into many domains. Made more precise, this is the Peierls argument for the existence of a phase transition in the two-dimensional Ising model. Similar arguments apply for other models with discrete symmetries.

For systems with a continuous symmetry, however, the energetics of domain walls is quite different. If we form a domain of size ℓ by insisting that the spins near the centre of the domain point in the opposite direction from those far away, then the intervening spins have a distance $O(\ell)$ over which to relax. Since they may do this in a continuous fashion, the relative angle between two neighbouring spins will be $O(1/\ell)$, and the energy density of such a configuration is $O((1/\ell)^2)$. This yields a total energy $O(\ell^{d-2})$ for a domain whose volume is $O(\ell^d)$, as compared with $O(\ell^{d-1})$ in the discrete case. This means that the entropic effects will always win for $d \leq 2$.

This heuristic argument is corroborated by the Mermin–Wagner–Hohenberg theorem (which in the field theory literature is also called Coleman's theorem) which states that there cannot be any spontaneous breaking of a continuous symmetry in $d \leq 2$ dimensions. To gain an idea of the proof, consider the $O(n)$ model introduced on p.22, with degrees of freedom $\mathbf{s}(r)$, where $\mathbf{s}(r)^2 = 1$. Suppose that there is a phase with a spontaneous magnetisation. Then, at sufficiently low temperature, we would expect the spins all to point roughly in the same direction, say, along the 1-axis. So we may write $\mathbf{s} = (\sqrt{1 - \sigma^2}, \sigma)$, where σ has $n - 1$ components and $|\sigma| \ll 1$. In terms of this parametrisation, the reduced hamiltonian is

$$\mathcal{H} = -\tfrac{1}{2}\beta \sum_{r,r'} J(r-r')\sigma(r)\cdot\sigma(r') + \cdots = \text{const.} + \tfrac{1}{2}K \int (\nabla\sigma)^2 d^d r + \cdots$$

$$(6.1)$$

using a continuum notation. The neglected terms are higher order in σ and therefore small by hypothesis. The degrees of freedom σ

are called, in the language of magnetism, *spin waves*. More generally, for any spontaneously broken continuous symmetry, these *Goldstone modes* should exist. Notice that, at least to lowest order, there is no term proportional to σ^2 in $\mathcal{H}$, and therefore these excitations have infinite correlation length, as was found in the mean field approximation in Section 2.2. Goldstone's theorem asserts in general that, whenever a continuous symmetry is truly broken, the inverse correlation length for these modes remains zero even when all the higher order interactions are taken into account. To lowest order, the two-point function of these modes is given by Gaussian integration (see Appendix)

$$\langle \sigma_a(r)\sigma_b(0) \rangle = \frac{\delta_{ab}}{K} \int_{\mathrm{BZ}} \frac{e^{ik\cdot r}}{k^2} \frac{d^d k}{(2\pi)^d}. \tag{6.2}$$

Let us now calculate the first order correction to the spontaneous magnetisation $\langle s_1(r) \rangle = 1 - \frac{1}{2}\langle \sigma(r)^2 \rangle + \cdots$. The correction is proportional to

$$\frac{(n-1)}{K} \int_{\mathrm{BZ}} \frac{d^d k}{k^2}. \tag{6.3}$$

where the integral is over the first Brillouin zone. (Strictly speaking, the k^2 form of the denominator is not valid near the zone boundary, but, as will be seen, this does not compromise the argument.) For $d > 2$, the integral is finite and gives the first order term in the low temperature expansion of the magnetisation, which, at sufficiently low temperatures ($K^{-1} \ll 1$) is much smaller than the leading term.

For $d \leq 2$, however, the integral is divergent at $k = 0$, so that the expansion makes no sense. Therefore the original assumption of the existence of an ordered phase for non-zero temperature must be incorrect. It is not difficult to refine this argument to construct a rigorous proof of the theorem. Physically, we may say that, for $d \leq 2$, the fluctuations around any potentially ordered phase are always sufficiently strong as to destroy it.

6.2 The two-dimensional XY model

The case $n = 2$, $d = 2$ of the $O(n)$ model is particularly interesting. Although the Mermin–Wagner–Hohenberg theorem ensures that it cannot have an ordered phase, it nevertheless undergoes a phase transition of a novel kind, due to the unbinding of defects. When $n = 2$, it is convenient to parametrise the components of the spin by $s(r) = (\cos\theta(r), \sin\theta(r))$, where $\theta(r)$ is an angle. This model has important physical applications, apart from the obvious one to XY magnets. For example, the angle $\theta(r)$ may be taken to represent the phase of the macroscopic wave function in liquid helium. The ordering transition of the XY model then turns out to correspond to the transition to the superfluid state. In addition, as we shall see, the two-dimensional XY model turns out to be related to models of the *roughening transition* of crystalline surfaces. The mechanism of the transition in the XY model is similar to that responsible for melting of two-dimensional lattices in some circumstances.

In terms of the angular degrees of freedom, the reduced hamiltonian for an XY magnet has the form

$$\mathcal{H} = -\tfrac{1}{2}\beta \sum_{r,r'} J(r - r') \cos(\theta(r) - \theta(r')). \qquad (6.4)$$

At low temperatures, we expect the argument of the cosine to be small, so we may expand in powers of $(\theta(r) - \theta(r'))$. Since θ is a continuous variable, there is no difficulty, at least formally, in going to the continuum limit, whence

$$\mathcal{H} = \text{const.} + K \int [\tfrac{1}{2}a^2(\nabla\theta)^2 - u_4 a^4(\nabla\theta)^4 + \cdots]\frac{d^2r}{a^2}, \qquad (6.5)$$

where $K = \beta R^2 J/2a^2$, in the same notation as Chapter 2. The coefficients u_4, *etc.*, of the $(\nabla\theta)^4$ and higher order terms all come with positive powers of a, and therefore, just as in Section 5.5, they are irrelevant at the Gaussian fixed point. Thus the low-temperature properties of the model are controlled by this Gaussian fixed point. However, unlike the Gaussian fixed point considered in connection with the ϵ-expansion in Section 5.5, we are not free to rescale θ so that the coefficient K is unity. This is because θ is an angular variable, with a definite periodicity of 2π. Therefore K is not a redundant variable. However, from (6.5), it is dimensionless, and therefore is unaffected by any rescaling of the lattice spacing a. Since there are no higher order interaction terms which might modify this result, we conclude that K is *exactly marginal*, that is, it parametrises a *line* of fixed points.

Let us work out some of the properties of these fixed point models. The correlation function $\langle\theta(r_1)\theta(r_2)\rangle$ of the angles is given, by the methods of Gaussian integration (see Appendix), as

$$G(r_1 - r_2) = \frac{1}{K}\int_{\text{BZ}} \frac{e^{ik\cdot(r_1-r_2)}}{k^2}\frac{d^2k}{(2\pi)^2}. \tag{6.6}$$

This is infrared divergent at $k = 0$. However, it turns out that all physical correlation functions are given in terms of

$$G(r_{12}) - G(0) = -\frac{1}{K}\int_{\text{BZ}} \frac{1 - e^{ik\cdot(r_1-r_2)}}{k^2}\frac{d^2k}{(2\pi)^2}, \tag{6.7}$$

which is finite. We shall need its behaviour as $r_{12} = |r_1 - r_2| \to \infty$. The easiest way of finding this is to note that, on dimensional grounds, the right hand side is a function of only the ratio (r_{12}/a), so that we may equally well extract the leading behaviour as $a \to 0$. To do this we may replace the Brillouin zone by a cut off $|k| < 1/a$. As $a^{-1} \to \infty$, the divergent part of the integral goes like $\int^{1/a}(dk/k) \sim -\ln a$, so that

$$G(r) - G(0) \sim -\frac{1}{2\pi K}\ln(r/a) + (C/K) + \cdots, \tag{6.8}$$

where C is a lattice-dependent constant of order unity. The coefficient of the logarithm is, however, universal.

More interesting is the behaviour of the spin-spin correlation function $\langle \mathbf{s}(r_1)\cdot\mathbf{s}(r_2)\rangle = \text{Re}\langle e^{i(\theta(r_1)-\theta(r_2))}\rangle$. By the rules of Gaussian integration explained in the Appendix, this is

$$e^{-\frac{1}{2}\langle(\theta(r_1)-\theta(r_2))^2\rangle} = e^{-(G(0)-G(r_{12}))}. \tag{6.9}$$

On substituting (6.8) we then find that, as $r_{12} \to \infty$,

$$\langle \mathbf{s}(r_1)\cdot\mathbf{s}(r_2)\rangle \sim \frac{\text{const.}}{r_{12}^{1/2\pi K}}. \tag{6.10}$$

The correlation function therefore decays with a power law, as expected at a fixed point, but the exponent $\eta = 1/(2\pi K)$ is non-universal and *continuously varying* along the line of fixed points parametrised by K. The fact that the correlation function approaches zero is consistent with the absence of any spontaneous magnetisation, as implied by the Mermin–Wagner–Hohenberg theorem. This low temperature phase is not the usual paramagnetic one expected at high temperatures, since the correlation function

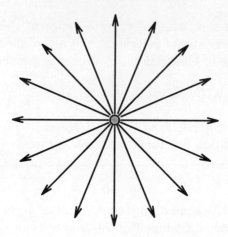

Figure 6.1. A single vortex configuration.

decays as a power law rather than exponentially. Such a phase is said to have *quasi-long range order*.

However, this state of affairs cannot persist up to arbitrarily high temperatures, since it is known that exponential decay of the correlation functions must occur for sufficiently large T (that is, small K). This inadequacy of the analysis so far is traceable to the neglect of the fact that θ is a periodic variable which should be identified modulo 2π. The significance of this is that low energy configurations exist in which θ varies very little between neighbouring sites, but where, if we follow its behaviour around a large closed curve, θ may in fact change by some nonzero multiple of 2π. Such configurations are called (by analogy with the case of superfluidity) *vortices*. An example is shown in Figure 6.1. In general, the angle θ may change by any integer multiple of 2π in winding around the core of such a vortex. This is called the vorticity of the configuration. From symmetry, the configuration with lowest energy and given vorticity has the form $\theta \sim n\phi$ far from the core, where ϕ is the polar angle. Its energy is $E = \frac{1}{2}K \int_a^L (n/r)^2 d^2r \sim \pi n^2 K \ln(L/a)$, where L is the linear size of the system. In the thermodynamic limit, therefore, single vortices have infinite energy and should not appear. However, it turns out that multi-vortex configurations have finite energy as long as the total vorticity vanishes.

There is a simple argument, known as the *Kosterlitz–Thouless criterion*, for determining when these vortex configurations begin to be important. Suppose they are dilute with a typical separation L_0. The energy of each vortex will then be given approximately by the above result with L replaced by L_0. However, such a vortex may be in roughly $(L_0/a)^2$ different positions, so it has an entropy $\sim 2\ln(L_0/a)$. Its reduced free energy is therefore $(\pi n^2 K - 2)\ln(L_0/a)$ (recall that K already contains a factor $1/k_B T$). For $K > 2/\pi$, this is always positive and the occurrence of free vortices is therefore suppressed. For $K < 2/\pi$, however, vortices with $n = \pm 1$ may proliferate. Once they do so, the above calculation based on the Gaussian approximation becomes unreliable. However, this argument does indicate that, by the time the temperature T has reached $T_{KT} = \pi J R^2/4 k_B a^2$, the vortices will have proliferated. Once they do so, as we shall see, a finite correlation length is generated. The Kosterlitz–Thouless criterion therefore gives an upper bound on the critical temperature of the model.

6.3 The solid-on-solid model

Before turning to the renormalization group analysis of the two-dimensional XY model, let us briefly describe the connection to the equally important solid-on-solid (SOS) model of surface roughening. This is through a *duality transformation* of a type very common in the study of two-dimensional lattice models. For definiteness, we work on a square lattice.

Although the original formulation of the XY model leads to a cosine interaction of the form (6.4) between neighbouring angular variables $\theta(r)$, the only important features of this form are that it reproduces, in the low temperature limit, the Gaussian model (6.5), and that it is suitably periodic in θ. A more general choice would be to substitute for the Boltzmann weight corresponding to the bond connecting the nearest neighbour sites r_i and r_j

$$e^{J\cos(\theta(r_i)-\theta(r_j))} \rightarrow \sum_{m_{ij}=-\infty}^{\infty} e^{-\tilde{J}(m_{ij})+im_{ij}(\theta(r_i)-\theta(r_j))}, \qquad (6.11)$$

where m_{ij} is an integer, and $\tilde{J}(m) = \tilde{J}(-m) \propto m^2$ as $|m| \to \infty$. Such a choice naturally has the above properties. If this expansion

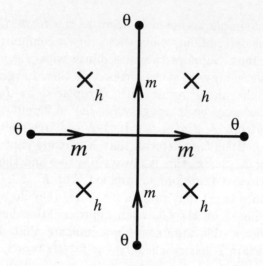

Figure 6.2. Duality transformation in the XY model.

is inserted, for every nearest neighbour bond, into the partition function, the traces over the angles θ may be carried out explicitly. At a given site i, this has the form (see Figure 6.2)

$$\int_0^{2\pi} e^{i\sum_j m_{ij}\theta(r_i)} \frac{d\theta(r_i)}{2\pi} = \delta_{\sum_j m_{ij},0}. \qquad (6.12)$$

At each site there is therefore a constraint that the sum of the m-variables on each bond entering the site should vanish. It is useful to think of these variables as components of a vector field (m_x, m_y) with m_x defined on the bonds in the x-direction, and m_y on those in the y-direction. In this language, the constraint (6.12) implies that this lattice vector field is divergence-free. As in the continuum, it may therefore be written as the curl of a scalar field h: $m_\mu = \sum_\nu \epsilon_{\mu\nu}\Delta_\nu h$, where $\epsilon_{\mu\nu}$ is the totally antisymmetric symbol and Δ_ν is a finite lattice difference. On the lattice, this amounts to defining integer valued variables $h(R)$ at the sites R of the *dual lattice*. This is the lattice with a site in each elementary plaquette, or square, of the original lattice. For the case of a square lattice, the dual lattice is also square. The variable m on a given bond is then parametrised in terms of the difference of the $h(R)$ on the dual sites on either side of the given bond, due attention being paid to the question of signs (see Figure 6.2).

The result of this transformation is that the partition function is now written as trace over the dual variables

$$Z = \text{Tr}_h e^{-\sum_{R,R'} \tilde{J}(h(R)-h(R'))}, \qquad (6.13)$$

where the sum is over nearest neighbours (R, R') on the dual lattice. (6.13) is a typical result of duality. The dual variables are Fourier conjugates to the original ones,[†] and the Boltzmann weights of the dual model are the Fourier components of those of the original. In particular, we see that when the function $\tilde{J}(m)$ is strongly peaked about $m = 0$, corresponding to low temperature in the dual model, the degrees of freedom in the original model are relatively unconstrained, corresponding to high temperature. Duality therefore generally maps low into high temperature and *vice versa*.[‡]

The physical interpretation of the dual model (6.13) is as follows. Imagine a simple cubic crystal with a surface oriented, on average, normal to the 001-axis. The height of the surface, neglecting overhangs, may be parametrised by an integer-valued height variable $h(R)$, giving the location on the z-axis of a portion of the surface above the point R in the xy-plane. The surface energy will be the sum of a constant piece, proportional to the projected area on the xy-plane, and a fluctuating piece due to the steps on the surface. As long as this energy is local, it may be expressed in the form which appears in (6.13). At low temperatures in the XY model, the Gaussian approximation (6.5) is good, so that small values of $\Delta\theta$ and, therefore, large values of Δh dominate the partition sum. In this phase the surface is *rough*: it is a reasonable approximation to treat the variables $h(R)$ as though they were continuous, and the hamiltonian as though it were a simple Gaussian. In that case $\langle (h(R) - h(0))^2 \rangle \sim \ln R$ as $R \to \infty$, so that eventually the height of the surface wanders far from its value at the origin. On the other hand, at low temperatures in the SOS model, all the heights are the same, and the surface is said to be smooth. In this phase, the discreteness of the height variables is

---

[†] In general, the dual variables label irreducible representations of the symmetry group of the original model, in this case $U(1)$.

[‡] For models which are *self-dual*, for example, the Ising model, this gives an exact way of determining the critical point, assuming there is only one transition.

relevant, and the system feels the effect of the underlying periodicity in the z-direction.

The fact that the XY model and the SOS model are relatively dual means that the singular parts of their free energies, and hence quantities like the specific heat, are simply related. However, the relation between the correlation functions of the two models is more complicated: in general, scaling operators which are local in one model are non-local, or multi-valued, in the other.

6.4 Renormalization group analysis

Since the line of fixed points of the XY model corresponds to a Gaussian model, the perturbative renormalization group analysis of Section 5.2 may be applied quite simply. Let us modify the Gaussian model of (6.5) by allowing a non-zero fugacity y_0 which controls the appearance of vortices. We shall not need an explicit expression for this term. In a particular model, we expect that the initial value of y_0 is of order unity, before renormalization. If we expand the partition function in powers of y_0, the first non-zero term will occur at $O(y_0^2)$, corresponding to a pair of vortices with equal and opposite vorticity of unit magnitude. The contribution of such configurations to the partition function will then be of the form

$$y_0^2 \int e^{-E(r_1, r_2)} \frac{d^2 r_1 d^2 r_2}{a^4}, \qquad (6.14)$$

where $E(r_1, r_2)$ is the reduced free energy of a pair of vortices fixed at locations r_1 and r_2, evaluated within the Gaussian model. At low temperatures, the configuration of lowest energy under these constraints corresponds to

$$\theta(r) = \Theta(r - r_1) - \Theta(r - r_2), \qquad (6.15)$$

where $\Theta(r - r_i)$ is the angle which the vector $r - r_i$ makes with, say, the positive x-axis. The energy $\frac{1}{2} K \int (\partial \theta)^2 d^2 r$ of this configuration is easy to estimate in the limit $r_{12}/a \to \infty$, along the lines of the calculation on p.115, which gives

$$E(r_1, r_2) \sim 2\pi K \ln(r_{12}/a) + 2\pi K \tilde{C} + \cdots, \qquad (6.16)$$

with $\tilde{C}$ a non-universal constant of order unity. It is determined by the structure of the vortices on short distance scales of the order of a, and is called the *core energy*. Note that, compared with (6.8), the inverse temperature K now appears in the numerator multiplying the logarithm.

From this result, substituted into (6.14), we learn two things. First, if we are to normalise the vortex-vortex correlation function in the conventional manner so that its coefficient is of order unity, we should absorb the factor $e^{-2\pi K \widetilde{C}}$ in the fugacity coefficient y_0^2, so that the effective expansion parameter is $y \equiv y_0 e^{-\pi K \widetilde{C}}$. More importantly, we see that the scaling dimension of a single vortex operator along the Gaussian line of fixed points is $x_V = \pi K$, so that its renormalization group eigenvalue is $y_V = 2 - \pi K$, and the renormalization group equation for y in the vicinity of the line of fixed points is therefore

$$dy/d\ell = (2 - \pi K)y + \cdots. \tag{6.17}$$

We see that, for $K > 2/\pi$, the vortices are irrelevant, while the opposite is true for $K < 2/\pi$, consistent with the Kosterlitz–Thouless criterion described above. However, there is more information to be gained. Let us denote $2 - \pi K$ by x, so that $x = 0$ is the Kosterlitz–Thouless point, and $x \propto (T - T_{\mathrm{KT}})$ when this is small. The variable x couples to the energy density $(\nabla \theta)^2$ in the Gaussian model. The form of (6.17) shows, by the analysis in Section 5.2, that the vortex operator must arise in the operator product expansion of itself with the energy density. Now, whenever this happens, it also follows that the energy density will appear in the operator product expansion of two vortex operators. This is a general result of conformal invariance, to be described in Section 11.2. In fact, if we are careful to normalise the operators correctly, conformal invariance implies that the corresponding operator product expansion coefficients are exactly equal. However, we shall not need to to do this here. The non-vanishing of this coefficient implies, however, that the renormalization group equation for the temperature has the form, near $x = y = 0$,

$$dx/d\ell = Ay^2 + \cdots, \tag{6.18}$$

where the coefficient A is of order unity. No further second order terms in (6.17) and (6.18) are allowed by the symmetry under $y \to -y$, which ensures that vortices may appear only in pairs.

Equations (6.17) and (6.18) are the *Kosterlitz renormalization*

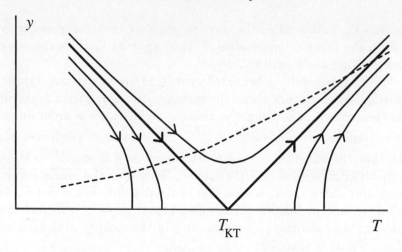

Figure 6.3. Kosterlitz–Thouless RG flows in the XY model.

group equations. The physical origin of the term $O(y^2)$ in (6.18) is due to *screening.* There is a very useful physical analogy between the vortices and electric charges of strength n, since they interact logarithmically, and, in two dimensions, this is a Coulomb potential. The quantity K may be thought of as the dielectric constant of the medium. Closely separated vortex pairs will be polarised by the fields of other, more distant vortices, and act to increase the effective dielectric constant, and therefore to screen their interaction. When, however, these close pairs are integrated out in the renormalization group by the change of short distance cut off a, this screening contribution to K will be lessened, and thus x will increase under renormalization. It is possible to compute this effect explicitly, but the calculation is quite lengthy. In any case, as we have seen, the form of the final result follows from very general principles. Note that vortices with vorticity $|n| > 1$ are irrelevant from the renormalization group point of view near $K = 2/\pi$, because of the additional factor of n^2 appearing in their scaling dimension.

　　The renormalization group flows near T_{KT} are shown schematically in Figure 6.3. There are two special trajectories of the form $y = Mx$. On substituting these into (6.17,6.18) we find $M = \pm A^{-1/2}$. The line $y = -A^{-1/2}x$ forms a separatrix which is in fact the critical surface. As the temperature T is increased, the line of points $y = y_0 e^{-\tilde{C}J/k_B T}$ corresponding to the initial model intersects this line at a value of x which corresponds to the critical temperature T_c. To see this, note that for $T < T_c$, the renormalization group trajectories ultimately flow into the Gaussian line of fixed points $y = 0$. Thus, the correlation functions exhibit quasi-long range order. On the other hand, for $T > T_c$ all trajectories ultimately flow to larger values of y. Although the expansion in powers of y breaks down in this region, we may argue that in this case the vortices are free to move, rather than existing only in bound pairs. This may be seen by computing the polarisability, in the language of the Coulomb gas. Suppose we modify the hamiltonian by adding an 'electric field'

$$\mathcal{H} \rightarrow \mathcal{H} + \mathbf{E} \cdot \sum_R \mathbf{R}\, n(R). \qquad (6.19)$$

Then the polarisability χ_E is the derivative $\partial^2 f / \partial E_i \partial E_i$ of the free energy, and is therefore given in terms of the vortex-vortex correlation function:

$$\chi_E = \sum_R R^2 \langle n(R) n(0) \rangle. \qquad (6.20)$$

To lowest order in y^2, this correlation function behaves as $R^{-2\pi K}$ (see (6.14)), and so, for small y, χ_E actually diverges when $K < 2/\pi$, that is $T > T_{\mathrm{KT}}$. Since all points just above the critical line $T = T_c(y_0)$ renormalize initially towards the region where y is small and $T > T_{\mathrm{KT}}$, it follows that χ_E will diverge at these points also, and, as increasing T should only increase the polarisability, we may argue that it should diverge throughout the entire region to the right of the separatrix. The significance of a divergent polarisability is that the charges (vortices) are free to move: in a small but finite **E**-field all the positive charges will move to one end of the system and all the negative charges to the other. In the absence of such a field, such a system is expected to exhibit *Debye screening*: a positive test charge will be completely surrounded by mobile negative charges in such a way that its effective electric field decays exponentially, over a distance scale given by the Debye screening length. This finite length may be related to the exponential fall off of the spin-spin correlation function in this phase, although the connection is not straightforward. This, then, is the

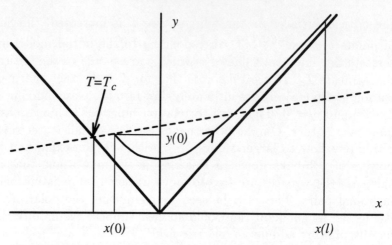

Figure 6.4. Blow-up of the critical region of Figure 6.3, showing a typical trajectory.

usual paramagnetic high temperature phase.

We may use the renormalization group to find how this length ξ diverges as $T \to T_c+$. To do this, we follow a particular renormalization group trajectory from just above T_c to a point where $x(\ell) = O(1)$, where we assume that the correlation length is some finite quantity ξ_0. This is illustrated in Figure 6.4. We first have to find the equation describing the trajectory. Dividing equations (6.17) and (6.18), we find $dy/dx = x/Ay$, which may be integrated to give $Ay^2 - x^2 = \text{const}$. The constant is determined by the initial data: $y = y(0)$, $x = x(0) \approx -A^{1/2}y(0)(1 + t)$, where $t = (T - T_c)/T_c$. This fixes the constant to be $\approx 2Ay(0)^2 t$. The value of the scale parameter ℓ when $x(\ell) = x(1)$ is then

$$\ell = \int_0^\ell d\ell' = \int_{x(0)}^{x(1)} \frac{dx}{Ay^2} = \int_{x(0)}^{x(1)} \frac{dx}{x^2 + 2Ay(0)^2 t}. \tag{6.21}$$

For small t the integral is insensitive to its limits and we may formally take $x(0) \to -\infty$ and $x(1) \to \infty$. Thus $\ell \sim \pi/(2Ay(0)^2 t)^{1/2}$. Finally, since the correlation length transforms trivially under the renormalization group, $\xi = \xi_0 e^\ell$, so that

$$\xi \propto e^{b/y(0)t^{1/2}}, \tag{6.22}$$

where b is a number of order unity. We see that in this case there is no exponent ν. Indeed, ξ diverges faster than any power of

t. Great care must be taken, however, in comparing (6.22) with data, experimental or numerical. In order to do this properly, it is necessary to estimate the size of the corrections to this result. The leading correction, however, is an additive constant in (6.21), which gives an overall multiplicative constant in (6.22). The true t-dependent corrections arise from yet higher order corrections to (6.21). A careful analysis shows that, for reasonable physical values of the parameters, the scaling region where (6.22) holds is of order $t < 10^{-2}$, and the corresponding correlation length is at least 10^8 lattice spacings! This makes such a prediction useful only for very pure experimental systems, and quite irrelevant for numerical simulations on finite systems.

The singular part of the free energy will scale as $e^{-2\ell} \sim \xi^{-2}$. This corresponds to a very weak singularity in the specific heat. All derivatives of the free energy are in fact continuous at T_c, so this is an infinite-order transition in the traditional classification. In fact, the main feature of the specific heat is expected to be a broad peak well above T_c corresponding to the proliferation of free vortices.

Below the critical temperature, the renormalization group trajectories flow into the Gaussian line of fixed points, at some particular value $K(\infty)$, which depends on T. The fact that the effective value of K is finite in this phase means, in the Coulomb gas analogy, that the screening effect is only partial, acting to increase the dielectric constant, but not driving it to infinity as occurs in the free vortex, Debye phase. The asymptotic behaviour of the spin-spin correlation function is the same as that in the Gaussian model at $K(\infty)$, so that $\eta = 1/2\pi K(\infty)$ is a non-universal, but monotonically increasing, function of T. However, at T_c, $K(\infty) = 2/\pi$, so that the value of $\eta(T = T_c) = \frac{1}{4}$ is universal.

The quantity $K(\infty)$ is related to an important observable of the XY model known as the *stiffness*. This is related to the sensitivity to the boundary conditions. Imagine a large system, of linear size $L \times L'$. In the y-direction, we impose periodic boundary conditions, but, in the x-direction, we consider either periodic boundary conditions $\theta(L, y) = \theta(0, y)$, or *twisted* boundary conditions $\theta(L, y) = \theta(0, y) + \alpha$. In the Gaussian approximation, where $\mathcal{H} = \frac{1}{2} K \int (\nabla\theta)^2 d^2 r$, the response of the reduced free energy to this change in the boundary conditions is straightforward to evaluate, because we can make the shift $\theta(x, y) \rightarrow \theta(x, y) - \alpha x/L$, which will bring $\mathcal{H}$ back to its periodic form, with an additional term $\frac{1}{2}K(\alpha/L)^2 LL'$. The change in the reduced free energy is therefore

$$\Delta f = \frac{1}{2}K(\alpha/L)^2. \qquad (6.23)$$

The coefficient of $(\alpha/L)^2$ in Δf is generally called the (spin wave) stiffness Υ. In the full theory, including vortices, the stiffness is given by (6.23) with K replaced by $K(\infty)$. This is because, under the renormalization group, the reduced free energy with a given twist $f(\alpha, K, y, L)$ transforms in a simple manner:

$$f(\alpha, K, y, L) = e^{-2\ell}f(\alpha, K(\ell), y(\ell), Le^\ell) + \text{inhomogeneous terms.} \qquad (6.24)$$

The inhomogeneous terms come from the function g in (3.25). Since this arises from tracing out the short wavelength degrees of freedom, however, it should be independent of α, so that these terms drop out in the difference Δf. In the low temperature phase, as $\ell \rightarrow \infty$, $y(\ell) \rightarrow 0$ and $K(\ell) \rightarrow K(\infty)$, giving the result. Note in particular that the stiffness will take on a universal value of $1/\pi$ at $T = T_c$. However, in the high temperature phase, the stiffness vanishes, because the free energy should exhibit exponential $e^{-L/\xi}$ dependence on boundary effects. When the XY system is used as a model for thin helium films, it turns out that the stiffness is related to the superfluid density ρ_s. A prediction of this theory is therefore of a jump in ρ_s at T_c. The universal value of Υ at the transition implies that $\hbar^2\rho_s(T_c)/m^2 k_B T_c = 2/\pi$, where m is the mass of a helium atom.

All the above analysis was based on the perturbative analysis of the renormalization group, and is justified only for small y. Since the initial value of $y \sim e^{-\widetilde{C}K}$, at low enough temperatures $K \gg 1$, this is surely valid. Whether it applies all the way up to the transition depends on the value of the core energy $\widetilde{C}$. As long as this is large enough, the above theory should apply. There have been suggestions, however, that if $\widetilde{C}$ is made sufficiently small, the transition to the high temperature phase is first order, rather than of the Kosterlitz–Thouless type described above. For helium films, however, the KT theory seems to be the correct description.

6.5 The $O(n)$ model in $2 + \epsilon$ dimensions

For $n > 2$ and $d > 2$, the $O(n)$ model turns out to have a conventional phase transition into a ferromagnetically ordered low temperature state, of the type suggested by mean field theory (Section 2.2) and the ϵ-expansion below four dimensions (Section 5.7). However, since the Mermin–Wagner–Hohenberg argument indicates that long range order in $d = 2$ is destroyed by fluctuations which are only logarithmically divergent, one might expect the critical temperature in $2 + \epsilon$ dimensions to approach zero as $d \to 2+$. In that case, the non-trivial critical fixed point is close to the trivial, zero temperature fixed point, so that the perturbative renormalization group may be applied.

We begin from a continuum version of the $O(n)$ hamiltonian

$$\mathcal{H} = \frac{1}{2T} \int (\nabla \mathbf{s})^2 \frac{d^d r}{a^{d-2}}, \qquad (6.25)$$

where $\mathbf{s}^2 = 1$. The factors of the lattice spacing a arise from replacing the finite lattice difference by a derivative, and from the volume a^d of the unit cell, in the usual way. Rather than measuring the exchange coupling in units of $k_B T$, we have measured the temperature T in units of the exchange coupling (divided by k_B), since we want to be able to take the zero temperature limit smoothly. In the above form, the $O(n)$ symmetry is explicit, but perturbation theory is not possible. Since we want to expand about the ordered state at $T = 0$, it is necessary first to break the symmetry, by parametrising $s_1 = \sqrt{1 - \sigma^2}$, $s_j = \sigma_j$ $(2 \leq j \leq n)$, as in Section 6.1 above. This leads to the hamiltonian

$$\mathcal{H} = \frac{1}{2T} \int \left[(\nabla \sigma)^2 + (\nabla \sqrt{1 - \sigma^2})^2 \right] \frac{d^d r}{a^{d-2}}. \qquad (6.26)$$

In this form, the model is known as the non-linear σ-model, since the full $O(n)$ symmetry is realised nonlinearly on the $n-1$ fields σ_j. It is now apparent that we may do perturbation theory in T after rescaling $\sigma \to T^{1/2}\sigma$, so that, to lowest order, the model is Gaussian. The interaction terms are found by expanding out the square root, so that the lowest order term is of the form $T \int (\nabla \sigma^2)^2 d^d r$. In order to implement the perturbative renormalization group, it is then necessary to find the coefficient of this term in its operator product expansion with itself! This rather technical exercise would take us beyond the scope of this book.† Fortunately, there is much simpler way of obtaining the first order renormalization group equations, which relies on what we have already learned of the case $n = 2$, $d = 2$.

The two-dimensional XY model was shown in Section 6.2 to exhibit a line of fixed points at low temperatures. This originated in the fact that the parametrisation $s_1 = \cos\theta$, $s_2 = \sin\theta$ leads to a Gaussian form for the hamiltonian, at least when vortices are excluded. This suggests that, for arbitrary $n \geq 2$, we parametrise

$$s_1 = \sqrt{1 - \mathbf{t}^2} \cos\theta \tag{6.27}$$

$$s_2 = \sqrt{1 - \mathbf{t}^2} \sin\theta \tag{6.28}$$

$$s_j = t_j \quad (2 < j \leq n). \tag{6.29}$$

In this form, the hamiltonian is

$$\mathcal{H} = \frac{1}{2T} \int [(1 - \mathbf{t}^2)(\nabla\theta)^2 + (\nabla\sqrt{1 - \mathbf{t}^2})^2 + (\nabla\mathbf{t})^2] \frac{d^d r}{a^{d-2}}. \tag{6.30}$$

Note that the phase angle θ enters very simply. The partition function is given by an integral of $e^{-\mathcal{H}}$ over both the θ and $\mathbf{t}$ degrees of freedom, and we may imagine performing the partial trace over $\mathbf{t}$, leaving an effective hamiltonian depending only on θ. At low temperatures, this may be done order by order in T by imagining rescaling $\mathbf{t} \to T^{1/2}\mathbf{t}$, and expanding the hamiltonian as before. To first order, the result is very simple: we replace the fluctuating quantity $\mathbf{t}^2$ by its average $\langle \mathbf{t}^2 \rangle$. This is straightforward to work out using the rules of Gaussian integration (see Appendix). To lowest order in T,

$$\langle t_i(r) t_j(0) \rangle = T a^{d-2} \delta_{ij} \int \frac{e^{ik \cdot r}}{k^2} \frac{d^d k}{(2\pi)^d}, \tag{6.31}$$

whence $\langle \mathbf{t}^2 \rangle \sim (n - 2)T/(2\pi\epsilon) + O(T^2)$, to first order in $\epsilon \equiv d - 2$. Exactly in $d = 2$, this gets replaced by $\langle \mathbf{t}^2 \rangle \sim (n-2)(T/2\pi)\ln a^{-1}$, since now the integral must be cut off. The all-important factor of $(n-2)$ arises because this is this number of degrees of freedom t_j.

For $d = 2$, and to first order in T, the effective hamiltonian for the θ degrees of freedom is therefore that of the XY model, with an effective coupling (*cf.* (6.5))

$$K_{\text{eff}} = T^{-1} - \frac{n - 2}{2\pi} \ln a^{-1} + O(T). \tag{6.32}$$

† There is also a tricky question of field renormalization to be addressed.

However, we know that in the XY model the coupling K is not renormalized at low temperatures (except by vortices which are non-perturbative effects). Thus, under a rescaling $a \to ae^{\ell}$, $dK_{\text{eff}}/d\ell = 0$. Substituting in the right hand side of (6.32) then gives the first order renormalization group equation $dT/d\ell = (n - 2)T^2/2\pi + O(T^3)$ in $d = 2$. The generalisation to $d = 2 + \epsilon$ is straightforward: dimensional analysis of (6.30) shows that the renormalization group eigenvalue of T at the $T = 0$ fixed point is simply $-\epsilon$. Therefore the full renormalization group equation is

$$dT/d\ell = -\epsilon T + \frac{n-2}{2\pi}T^2 + O(T^3, \epsilon T^2). \qquad (6.33)$$

The form of this equation should be compared with that for the renormalization of the S^4 coupling in 4-ε dimensions, see (5.58). They differ in the signs of the two leading terms on the right hand side. As a result, the stability of the fixed points is exchanged. The Gaussian fixed point at $T = 0$ is now stable, while the non-trivial fixed point at $T^* = 2\pi\epsilon/(n - 2) + O(\epsilon^2)$ is unstable. The long distance behaviour in the two regions $T < T^*$ and $T > T^*$ is therefore controlled by different stable fixed points, and therefore T^* should be identified with the critical temperature T_c. The $T < T_c$ phase is ferromagnetic, since the flows end up at the zero temperature fixed point, where we know there is a non-zero spontaneous magnetisation. For $T > T_c$, the flows leave the range of validity of the low temperature expansion. The mean field analysis of this model suggests that there should be only one phase transition, consistent with the simplest assumption that these flows end up at the infinite temperature fixed point, and that this phase is paramagnetic.†

The difference between (6.33) and the analogous equation for the S^4 coupling in (5.58), despite the fact that they refer to the same model, lies in the physical interpretation of these equations. In the latter case, the equation describes flows *within* the critical surface, with respect to which the critical fixed point should be stable. On the other hand, (6.33) describes flows along the temperature axis, with respect to which the critical fixed point is unstable. The S^4 coupling in the continuous spin $O(n)$ model

---

† All this applies only to the case $n > 2$. As discussed in Section 9.3, it is possible to define this model for fractional $n < 2$, when the renormalization group flows are rather different.

considered in Section 5.7 controls the fluctuations in the length of the spins $s(r)$, which are frozen out in the non-linear sigma model of this section. Just as in the Ising model, these longitudinal fluctuations play an increasingly important role as the dimension is increased, ultimately leading to the Gaussian results for $d > 4$.

The non-trivial fixed point at $T = T^*$ has a relevant scaling variable $T - T^*$ with eigenvalue $\epsilon + O(\epsilon^2)$. This is to be identified with the thermal eigenvalue y_t. Thus, using the scaling relation (3.50), we find that

$$\boxed{\nu = 1/\epsilon + O(1)} \qquad (6.34)$$

Note that this is consistent with the result of the large n limit discussed in Section 5.7. To calculate the magnetic exponent, it is possible to derive the renormalization group equation for an external magnetic field h through the operator product expansion. However, a simpler way to obtain the lowest order result is to note that, in perturbation theory in T, the spontaneous magnetisation is

$$M = \langle \sqrt{1 - \sigma^2} \rangle \sim 1 - \tfrac{1}{2}\langle \sigma^2 \rangle = 1 - \frac{(n-1)T}{4\pi\epsilon} + O(T^2). \quad (6.35)$$

The last result follows from a similar calculation to that used above for $\langle \mathbf{t}^2 \rangle$. This low temperature expansion fails as $T \to T_c$, where we expect that $M(T) \propto (1 - T/T_c)^\beta$. If we assume that this form is valid (to leading order in ϵ) for all $T < T_c$, we may compare with the perturbative expansion and conclude that $\beta/T_c \sim (n-1)/4\pi\epsilon$, so that

$$\boxed{\beta = \frac{n-1}{2(n-2)} + O(\epsilon)} \qquad (6.36)$$

This implies from (3.31) that the magnetic eigenvalue is $y_h = d - (n-1)\epsilon/2(n-2) + O(\epsilon^2)$, which may then be used to reconstruct the renormalization group equation for h *a posteriori*:

$$dh/d\ell = h(d - (n-1)(T/4\pi) + O(T^2)). \qquad (6.37)$$

Note that the eigenvalue at $T = 0$ is $y_h = d$, as required for a discontinuity fixed point (Section 4.1). This result may also be checked in the case $n = 2$, where it leads to the scaling dimension $x = T/4\pi$, consistent with the result $\eta = 1/2\pi K$ of Section 6.2.

At the lower critical dimension $d = 2$, there is no critical fixed point. Instead the renormalization group trajectories flow out of the region of applicability of perturbation theory. The most economical assumption is that this is the paramagnetic phase with a finite correlation length $\xi(T)$. Although the perturbative renormalization group cannot be used to calculate this, it does predict its behaviour at low temperature, by use of the transformation law $\xi(T) = e^\ell \xi(T(\ell))$. Integrating the equation for $T(\ell)$ gives

$$T(\ell)^{-1} = -\frac{n-2}{2\pi}\ell + T^{-1}, \qquad (6.38)$$

so that, if we choose $T(\ell)$ so that $\xi(T(\ell)) = O(1)$, we find

$$\xi(T) \propto T^{\beta_2} e^{2\pi/(n-2)T}. \qquad (6.39)$$

The exponent β_2 of the prefactor on the right hand side comes from the $O(T^3)$ term in the beta function. Its precise value needs a more sophisticated analysis. As with the XY model, this rapid divergence of the correlation length corresponds to a very weak singularity in the free energy as $T \to 0$.

The non-linear sigma model in two dimensions is an important example of a field theory which is *asymptotically free*, that is its short distance behaviour (not the limit relevant for phase transitions) is governed by the weak coupling fixed point at $T = 0$. It also exhibits *dimensional transmutation*, that is, out of a hamiltonian with only a dimensionless parameter T (and a microscopic cut off a), the theory generates a macroscopic correlation length. Since both of these features are believed to be shared with non-Abelian gauge theories in four dimensions, the model has received a fair amount of attention in the field theory literature.

Exercises

6.1 Calculate the two-point function, and hence the scaling dimension and the renormalization group eigenvalue of the operator $\cos p\theta$ in the Gaussian model in two dimensions. Show that it is irrelevant above some critical temperature T_p. What is the nature of the low temperature phase $T < T_p$? Now consider such a perturbation on the full XY model, with vortices included. Show that for sufficiently large p, the renormalization group implies that this model should have two phase transitions as the temperature is varied.

6.2   Suppose that the angular degree of freedom of the XY model
      is allowed to take only those values which are multiples of
      $2\pi/p$, for some integer p. This is called the p-state clock
      model, for obvious reasons. Modify the duality transforma-
      tion of Section 6.3 to this case, and show that the dual model
      is also a p-state clock model with a different interaction. Con-
      sider in particular the case $p = 2$ and show that this reduces
      to the Ising model with a suitable change of variables. De-
      duce that the Ising model on a square lattice is self-dual, and
      find the critical value of the reduced coupling.

6.3   What is the renormalization group equation for the reduced
      coupling K of the Gaussian model in $2 + \epsilon$ dimensions? As-
      suming that the renormalization group equations for the XY
      model in $2 + \epsilon$ dimensions are simple deformations of those
      found for $d = 2$ in Section 6.4 (without worrying what is the
      physical interpretation of y in this case), show that, to first
      order in ϵ, there is a simple critical fixed point, and calculate
      the critical exponent ν there, to leading order.

6.4   Show that the renormalization group equation (6.37) for the
      magnetic field in the $O(n)$ model may also be derived as fol-
      lows. Suppose h couples to the component $s_1 = \sqrt{1 - \mathbf{t}^2}\cos\theta$.
      To the order required, this is equivalent to a magnetic field
      $h_{\text{eff}} = (1 - \frac{1}{2}\langle\mathbf{t}^2\rangle)h$ in the XY model. Then use the fact that,
      from (6.10), $dh_{\text{eff}}/d\ell = (d - K_{\text{eff}}^{-1}/4\pi)h_{\text{eff}}$. Construct an anal-
      ogous argument to derive the renormalization group equation
      for the field coupling to the operator $\sum_{i=1}^{n} s_i^4$ which describes
      cubic symmetry breaking in this model.

6.5   By considering the general form of the renormalization group
      equation for the temperature, show that the leading order re-
      sult $\nu \propto 1/\epsilon$ is generic just above the lower critical dimension,
      $d = d_l + \epsilon$, for any system.

6.6   By generalising the discussion on p.125 to arbitrary dimen-
      sion d, show that the stiffness coefficient should vanish as
      $(T_c - T)^{(d-2)\nu}$ as $T \to T_c-$. This is known as Josephson scal-
      ing.

7

Surface critical behaviour

One of the most striking aspects of critical behaviour is that of the crucial role played by the geometry of the system. Critical exponents depend in a non-trivial manner on the dimensionality d. The very existence of a phase transition depends on the way in which the infinite volume limit is taken, as discussed in Section 4.4. This happens because the critical fluctuations, which determine the universal properties, occur at long wavelengths and are therefore very sensitive to the large scale geometry. By contrast, the non-critical properties of a system are sensitive to fluctuations on the scale of the correlation length and are therefore much less influenced. This line of reasoning also suggests that not all points in a system are equivalent in the way the local degrees of freedom couple to these critical fluctuations. So far, we have considered the behaviour of scaling operators only at points deep inside the bulk of a system. Near a boundary, however, the local environment of a given degree of freedom is different, and we might expect to find different critical properties there. In general, such differences should extend into the bulk only over distances of the order of the bulk correlation length. However, at a continuous bulk phase transition, this distance diverges, and we should expect the influence of boundaries to be more pronounced.

The simplest modification of the bulk geometry to consider is that of a $(d-1)$-dimensional hyperplane bounding a semi-infinite d-dimensional system. Such a situation might be realised in nature by the two-dimensional surface of a crystal. Such a real surface will inevitably contain irregularities, However, as long as the typical size of these is less than the correlation length, under most circumstances it is appropriate to regard the boundary as planar.

7.1 Mean field theory

Before studying the effect of the fluctuations through the renormalization group, it is important to understand what differences appear already at the level of mean field theory near a boundary. For simplicity, we consider the example of a ferromagnetic Ising model. In mean field theory, the local magnetisation satisfies the equation (2.15)

$$M(r) = \tanh\left(\beta \sum_{r'} J(r, r')M(r')\right). \tag{7.1}$$

The mean field transition in the bulk occurs when $\beta \sum_{r'} J(r, r') = 1$, and the spontaneous magnetisation vanishes when this quantity is less than unity. Near the boundary, since there are fewer neighbours, we expect $\sum_{r'} J(r, r')$ to be less than its bulk value (unless for some reason the surface couplings are enhanced) so that the magnetisation near the boundary will be suppressed relative to its value in the bulk. The phase transition that happens on the boundary is then termed the *ordinary transition*. This may be quantified as follows. Close to the bulk critical point, we expect $M(z)$ to be small and to depend slowly on the distance z from the boundary. Making a Taylor expansion of M we then find

$$\sum_{r'} J(r, r')M(r') = J(z)M(z) + \tfrac{1}{2}R^2 J \frac{\partial^2 M(z)}{\partial z^2} + \cdots, \tag{7.2}$$

where $J(z) = \sum_{r'} J(r, r')$, and J and R^2 were defined in Chapter 2. The sum is restricted to $z' \geq 0$. If the interactions are short range, $J(z)$ is equal to J except in a narrow region near the boundary, so that we may approximate, in the continuum limit, $J(z) \sim J(1 - (R^2/2\lambda)\delta(z))$, thus defining the positive quantity λ which has the dimensions of length. The magnetisation then satisfies the ordinary differential equation

$$\tfrac{1}{2}R^2\frac{\partial^2 M}{\partial z^2} + M - (R^2/\lambda)\delta(z)M = (\beta J)^{-1}\tanh^{-1} M \tag{7.3}$$

$$\sim (\beta J)^{-1}(M + \tfrac{1}{3}M^3 + \cdots),$$

with the boundary condition that $M = 0$ for $z < 0$. This may be rewritten, near the bulk critical point $\beta J = 1$, as

$$\tfrac{1}{2}R^2\frac{\partial^2 M}{\partial z^2} = tM + \tfrac{1}{3}M^3, \tag{7.4}$$

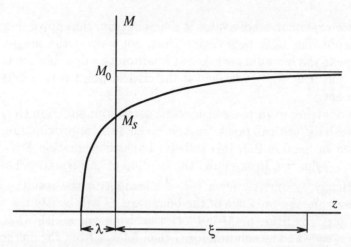

Figure 7.1. Behaviour of the magnetisation near the boundary, for $T < T_c$ in the case of the ordinary transition.

for $z > 0$, where as usual $t = (T - T_c)/T_c$, supplemented by the boundary condition $\partial M/\partial z = \lambda^{-1} M$ at $z = 0+$. If (7.4) were continued into $z < 0$, $M(z)$ would thus vanish at $z \approx -\lambda$. This distance, which is of microscopic size, is called the *extrapolation length*.

For $T > T_c$, the only solution has $M = 0$, as expected. Below the bulk critical temperature, M tends to its bulk value $\propto (-t)^{1/2}$ as $z \to \infty$, consistent with the usual mean field bulk magnetisation exponent $\beta = \frac{1}{2}$. Linearising the equation about this asymptotic value, we see that it approaches this value exponentially fast, over a distance scale given by the *bulk* correlation length $\xi \propto |t|^{-1/2}$, as shown in Figure 7.1. The solution of (7.4), which satisfies the condition that it vanish when extrapolated to $z = -\lambda$, has the scaling form

$$M(z) = (-t)^{1/2} f\left(\frac{z + \lambda}{Rt^{-1/2}}\right), \qquad (7.5)$$

where, as stated above, $f(X)$ approaches a constant as $X \to \infty$, and $f(0) = 0$. The magnetisation at or near the boundary $z = 0$ has a different temperature dependence, however. Since f is an analytic function of its argument (being the solution of a non-singular differential equation), it follows that $f(\lambda/Rt^{-1/2}) = O(t^{1/2})$, so that $M(0) \propto (-t)^{\beta_1}$ with $\beta_1 = 1$. β_1 is an example of a *surface exponent*, whose value is different from that governing the corresponding bulk behaviour. Thus, not only is the magnetisation near the boundary weakened in absolute value relative to the bulk, but it actually vanishes at the critical point with a different exponent.

As a further example, consider the correlation function $G(r_1, r_2)$ at the bulk critical point. In the mean field approximation described in Section 2.3, this satisfies Laplace's equation $\nabla^2 G = 0$ for $|r_1 - r_2|$ large. In the bulk, the solution of this equation has the spherically symmetric form $r_{12}^{-(d-2)}$, leading to the result $\eta = 0$. However, in the presence of the boundary, G will satisfy the same boundary condition as M in (7.4), that is, it will vanish whenever z_1 or $z_2 = -\lambda$. The solution for G then follows from the method of images, since we may think of it as the electrostatic potential at r_2 due to a charge at r_1, in the presence of a conducting boundary at $z = -\lambda$. Writing $\mathbf{r} = (\mathbf{x}, z)$,

$$G(r_1, r_2) \sim \frac{1}{\left((\mathbf{x}_1 - \mathbf{x}_2)^2 + (z_1 - z_2)^2\right)^{d/2-1}}$$
$$- \frac{1}{\left((\mathbf{x}_1 - \mathbf{x}_2)^2 + (z_1 + z_2 + 2\lambda)^2\right)^{d/2-1}}. \quad (7.6)$$

This result contains a number of interesting special cases. Far from the boundary, when both z_1 and z_2 are much greater than $|\mathbf{x}_1 - \mathbf{x}_2|$, we recover the bulk behaviour $r_{12}^{-(d-2)}$. If r_1 is close to the boundary, but r_2 is far away, the correlation function scales like $r_{12}^{-(d-1)}$, multiplied by a function of the angle which $\mathbf{r}_{12}$ makes with the normal to the surface. Finally, if both points are on or near the boundary, the asymptotic behaviour is proportional to r_{12}^{-d}. These results are consistent with the usual assignment of a scaling dimension of $x_h = d/2 - 1$ to the magnetisation operator in the bulk, with, however, a different *boundary scaling dimension* $x_{h,s} = d/2$ for the magnetisation close to the boundary. In Section 7.3 we shall make this concept more precise.

7.2 The extraordinary and special transitions

So far, we have assumed that the exchange couplings near the boundary are not enhanced above their bulk values. Because of the deficit in the number of close neighbours near the boundary, the magnetisation is suppressed there, and ordering at the boundary takes place only in response to the establishment of a spontaneous bulk magnetisation below the bulk critical temperature. However, if, for some reason, the bulk couplings are sufficiently enhanced, the surface may actually order at a temperature higher than that of the bulk. Since at this temperature the correlations through the bulk decay exponentially, the effective interactions between the boundary degrees of freedom are short range, and this *surface transition* will be in the universality class of the $(d-1)$-dimensional system with short range interactions. (This assumes, of course, that $d - 1 > d_l$, the lower critical dimension.) As the temperature is lowered further, the bulk orders at the bulk critical temperature. Since the surface quantities are coupled to the bulk, they themselves may also exhibit singularities at this point, which is called the *extraordinary transition*. The surface thus undergoes two separate transitions as the temperature is reduced. Finally, as the surface enhancement is reduced, these two lines of critical points meet at a multicritical point called the *special transition*. The phase diagram is shown schematically in Figure 7.2. The extraordinary transition may also be realised, even without enhancement of the surface couplings, by adding an external ordering field (a magnetic field in the case of a spin system) which is confined to the vicinity of the boundary. In that case, there is non-zero magnetisation near the boundary, which becomes singular as the bulk orders. Such surface fields are important in the case of binary fluids, when they correspond to a difference in the affinities of the wall for each component of the fluid.

Within mean field theory, the extraordinary transition corresponds to the case of a negative extrapolation length λ. From (7.4) we then see that, above the bulk T_c, $M(z)$ will decay exponentially to zero away from the boundary, over a length scale given by the bulk correlation length. Exactly at the bulk critical temperature, it decays into the bulk as z^{-1}. It may be shown (see Ex. 7.1) that close to T_c the surface magnetisation has the form $M(0) \sim M_0 + \text{const}.|t|$. This gives the surface magnetisation exponent $\beta_1 = 1$ for the extraordinary transition within mean field theory. Finally, the special transition occurs when $\lambda^{-1} = 0$, so that the boundary condition is simply $\partial M/\partial z = 0$. In this case,

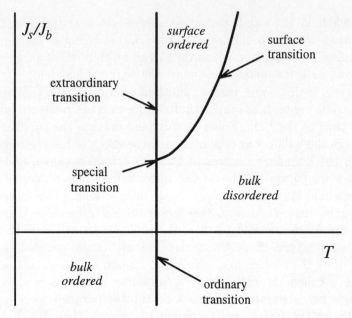

Figure 7.2. Phase diagram of a ferromagnet with enhanced surface couplings.

the only solution to (7.4) is to take $M(z)$ everywhere equal to its bulk value, so that, at least in mean field theory, the surface magnetisation exponents are equal to their corresponding bulk values.

7.3 Renormalization group approach

Once again, the simplest conceptual approach to the renormalization group study of surface critical phenomena is through the real space block spin transformations of the type described in Section 3.1. Since blocks near the boundary have a different number of neighbouring spins, the renormalization group transformation itself will be different there. Thus, even though we may begin with a system in which the couplings near the boundary are the same as those in the bulk, under renormalization this will no longer be the case. Instead, a new set of 'boundary couplings' $\{K_s\}$ will be generated. The renormalization group equations will then separate into two parts: the bulk transformation as in Section 3.3,

$$\{K'\} = \mathcal{R}(\{K\}), \tag{7.7}$$

and a transformation equation for the boundary couplings

$$\{K'_s\} = \mathcal{R}_s(\{K_s\}, \{K\}). \tag{7.8}$$

Since the boundary degrees of freedom interact through the bulk, their renormalization depends on the values of the bulk couplings, but not *vice versa*. In the same spirit as the central assumption that the bulk couplings $\{K\}$ remain of short range under renormalization, we also expect that the surface couplings $\{K_s\}$ are not only short range, but are also confined to some finite layer near the boundary.

An immediate consequence of the general structure of (7.7) and (7.8) is that, for each bulk fixed point $\{K^*\}$, there may be more than one fixed point $\{K^*_s\}$ for the surface couplings. Therefore, for a given bulk universality class, there may be several universality classes of surface transition. In the case of the ferromagnetic Ising model, we have already seen the possibility of three: the ordinary, extraordinary and special transitions. Linearising (7.8) about a given fixed point, we may define *boundary scaling variables*, in analogy to Section 3.8, which couple to *boundary scaling operators*. The corresponding renormalization group eigenvalues y_s are not necessarily the same as those in the bulk. The boundary scaling operators, which have boundary scaling dimensions x_s given by the generalisation of (3.55), namely

$$x_s = d - 1 - y_s, \tag{7.9}$$

are not necessarily directly identifiable with bulk scaling operators as they approach the boundary. In fact, the number of relevant variables at a boundary fixed point is, in general, less than that at the corresponding bulk fixed point, since the requirement of bulk criticality is already sufficient to guarantee that the surface is critical.

Once the scaling dimensions of the various boundary scaling operators are identified, we may read off various other exponents. For example, at the critical point, suppose the bulk and surface scaling dimensions of the magnetisation are x_h and $x_{h,s}$ respectively. Then the spin-spin correlation function should decay as $r^{-2x_{h,s}}$, $r^{-x_{h,s}-x_h}$ or r^{-2x_h}, depending on whether two, one or none of the points lie close to the surface. Conventionally, the corresponding exponents are denoted by $d - 2 + \eta_{\parallel}$, $d - 2 + \eta_{\perp}$

and $d - 2 + \eta$ respectively. The renormalization group immediately implies the scaling relation $\eta + \eta_\| = 2\eta_\perp$ between them. Similarly, the surface magnetisation below the bulk T_c should behave as $\xi^{-x_{h,s}}$ where $\xi \propto |t|^{-\nu}$ is the bulk correlation length. This leads to the scaling relation $\beta_1 = x_{h,s}/y_t$. Finally there are the exponents related to various types of magnetic susceptibility. The bulk susceptibility is proportional to $\int G(r_1, r_2) d^d r_2$ and therefore scales like ξ^{d-2x_h}. This gives the usual scaling relation (3.32) $\gamma = (d - 2x_h)/y_t$. On the other hand, we could consider the response of the boundary magnetisation to a change in the bulk magnetic field, which would be proportional to the same integral, with r_1 on the surface. This would then scale like $\xi^{d-x_{h,s}-x_h}$, yielding the exponent $\gamma_1 = (d - x_{h,s} - x_h)/y_t$. Finally, there is the response of the surface magnetisation to a change in a magnetic field confined to a layer near the boundary, which is proportional to $\int G(r_1, r_2) d^{d-1} r_2$, where both r_1 and r_2 are close to the boundary. This leads to the exponent $\gamma_{11} = (d - 1 - 2x_{h,s})/y_t$.

While the above scenario is compellingly attractive, it is clearly necessary to show that it is in fact realised under conditions where the various approximations are controlled. As an example, we consider the ferromagnetic Ising model in $4-\epsilon$ dimensions, which, in the bulk, is described by the continuous spin model (see Section 5.3)

$$\mathcal{H}_b = \int [(\nabla S)^2 + tS^2 + uS^4 + hS] d^d r. \tag{7.10}$$

In the above, the explicit factors of the microscopic distance scale a have not been included, so that S has dimension $(length)^{1-d/2}$. To this we now add a surface term, which is an $\int d^{d-1}x$ over an expression which may, in principle, include all possible powers of S and their local derivatives. However, very few of these are relevant at the Gaussian fixed point, namely

$$\mathcal{H}_s = \int [cS^2 + h_s^{(1)} S + h_s^{(2)} \partial_z S] d^{d-1} x. \tag{7.11}$$

Note the possibility of two surface symmetry breaking fields $h_s^{(1)}$ and $h_s^{(2)}$. The single derivative ∂_z is allowed in (7.11) because it does not integrate to zero, as it would in the bulk. At the Gaussian fixed point where $c = h_s^{(1)} = h_s^{(2)} = 0$, dimensional analysis shows that these scaling variables have renormalization group eigenvalues 1, $\frac{d}{2}$ and $\frac{d}{2} - 1$ respectively. Under renormalization c increases without bound and flows to ∞ or $-\infty$ according to its initial sign. In fact, c has the dimension of $(\text{length})^{-1}$ and may be identified with the inverse of the extrapolation length λ. The fixed point at $c = 0$ then corresponds to the multicritical point of the special transition, while the fixed point controlling the ordinary transition occurs at $c = +\infty$. Since the Boltzmann weights include a factor $\exp(-c \int S^2 d^{d-1}x)$, this enforces the boundary condition $S = 0$. This, in turn, means that the boundary field $h_s^{(1)}$ cannot couple at this fixed point, and the leading symmetry breaking field is therefore $h_s^{(2)}$. The physical reason for this curious result is as follows. In fact, the order parameter at the ordinary transition falls to zero at $z \approx -\lambda$, not exactly at the surface, but this is a microscopic distance which remains finite even at the bulk critical point. Thus, in the renormalized theory, it is effectively zero. To compute the actual order parameter on the surface we should note that $S(0) \approx \lambda \partial_z S|_{z=-\lambda}$, so that $\partial_z S$ is, in fact, the correct boundary operator in the renormalized theory.

This implies that the boundary scaling dimension of the magnetisation at the $u = 0$ fixed point is, from (7.9), $(d - 1) - (d/2 - 1) = d/2$, which is in agreement with the result of mean field theory. Scaling implies that the surface magnetisation should vanish as $T \to T_c$ as $\xi^{-d/2} \sim t^{d/4}$, which agrees with the mean field result $\beta_1 = 1$ at $d = 4$, as expected. For $d > 4$, u is once again a dangerous irrelevant variable which modifies this scaling argument.

What about the energy density near the boundary? At the $c = \infty$ fixed point, where $S = 0$ on the boundary, the leading scaling operator in the even subspace is $(\partial_z S)^2$, which has scaling dimension d. We conclude that energy-energy correlations along the boundary fall off as r_{12}^{-2d}, at least at the $u = 0$ fixed point. It turns out that this is a general result for the ordinary transition, valid also for $d < 4$.†

At the special transition, the symmetry breaking field is $h_s^{(1)}$, with eigenvalue $d/2$. This leads to $\beta_1 = (d - 1 - d/2)/2 = \frac{1}{2}$ in $d = 4$, once again in accord with mean field theory.

We now consider what happens below four bulk dimensions,

---

† This is because the leading even boundary operator is the stress tensor (see Section 11.3), which always has scaling dimension d.

when the bulk coupling u goes to a non-trivial fixed point of $O(\epsilon)$. In the bulk, as discussed in Section 5.5, the leading magnetic eigenvalue is not modified by the interaction to first order in ϵ. This may be traced to the fact that S does not appear in the operator product expansion of S and S^4, equation (5.30). However, near the boundary, this is modified. In the bulk, we found it convenient to redefine the variable u (see (5.26)) so that it actually couples to

$$:S^4: \equiv S^4 - 3\langle S^2 \rangle S^2. \tag{7.12}$$

This choice eliminated Wick contractions on the same point. To lowest order, $\langle S^2 \rangle$ is given by the limit of the 2-point correlation function $G(r_1, r_2)$ for $|r_1 - r_2| \sim a$. In the bulk, this is simply a constant, so that the second term in (7.12) may be absorbed in a redefinition of t. However, at the $c = \infty$ fixed point the correlation function near the boundary is given, as in (7.6), by the method of images:

$$G(r_1, r_2) = r_{12}^{-(d-2)} - r_{1\bar{2}}^{-(d-2)}, \tag{7.13}$$

where $r_{1\bar{2}}$ is the distance between r_1 and the image point of r_2 in the surface. (Note that we are normalising the correlation function as in Section 5.5.) As $r_1 \rightarrow r_2$, the second term becomes $(2z)^{-(d-2)}$. Hence we should write the uS^4 term in the bulk hamiltonian as

$$uS^4 = u:S^4: -3u(2z)^{-2}S^2 + \cdots, \tag{7.14}$$

where the neglected term corresponds to the spatially uniform shift in t already described, and the second term, where we have set $d = 4$ to leading order in ϵ, is new. Thus the S^4 term actually induces a position dependent S^2 coupling near the boundary. This term does have a non-trivial operator product expansion with S, and gives rise to a first order correction to the boundary magnetic exponents.

To evaluate this, let us return to the gas analogy of Section 5.2, and consider the renormalization of $h_s^{(2)}$ which couples to $\partial_z S$ on the boundary. This is illustrated in Figure 7.3. The required operator product is

$$\partial_z S(\mathbf{x}, z) \cdot S(\mathbf{x}', z')^2 \sim 2\partial_z G(\mathbf{x}, z; \mathbf{x}', z') S(\mathbf{x}', z'), \tag{7.15}$$

where the factor of 2 comes from the number of distinct contractions. The coefficient $\partial_z G$, evaluated at $z = 0$, is simply $4z'/((\mathbf{x} -$

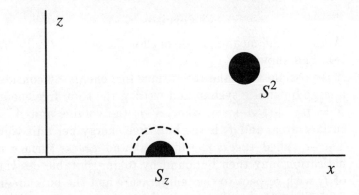

Figure 7.3. Boundary renormalization of $h_s^{(2)}$.

$\mathbf{x}')^2 + z'^2)^2$. In order to carry out the renormalization, $(\mathbf{x}', z')$ should be integrated over the interior of a hemispherical shell of radius a and thickness $a\delta\ell$, as shown by the dashed semicircle in Figure 7.3. Within this shell, we may approximate $S(\mathbf{x}', z')$ by $z'\partial_z S(\mathbf{x}, 0)$, since $S = 0$ on the boundary. These two factors of z' cancel the z'^{-2} coming from (7.14), so that we end up integrating $1/a^4$ over the half shell, which leads to a factor $\frac{1}{2}S_d$ as compared with S_d in the bulk. Absorbing this factor of S_d, as usual, into a redefinition of u then leads to the renormalization group equation

$$dh_s^{(2)}/d\ell = h_s^{(2)}(d/2 - 1 + 12u + O(u^2, \epsilon u)). \qquad (7.16)$$

At the Wilson–Fisher fixed point $u^* = \epsilon/72$ (see p.98), the boundary magnetic eigenvalue is $y_{h,s} = d/2 - 1 + \frac{\epsilon}{6} + O(\epsilon^2)$. Through the scaling relation $\beta_1 = (d - 1 - y_{h,s})\nu$ this finally gives $\beta_1 = 1 - \frac{\epsilon}{6} + O(\epsilon^2)$.

The magnetic exponents at the special transition are given, to first order in ϵ, by a simple variation of this calculation (see Ex. 7.6), since the effect of replacing the boundary condition $S = 0$ by $\partial_z S = 0$ is to change the relative sign of the second term in G in (7.6). Of course, this by no means exhausts the subject of boundary critical phenomena. In particular, the study of correlation functions near the boundary is of great interest, since these may be measured using X-ray scattering at grazing angles from crystal surfaces.

Exercises

7.1 Work out the mean field theory for the extraordinary transition and show that $\beta_1 = 1$.

7.2 It is possible to define the surface free energy by considering a large but finite system and writing the total free energy as $F = f_b V + f_s A + \cdots$, where V is the volume and A is the surface area, and f_s is the bulk free energy per unit volume. The so-called *excess* specific heat and *excess* surface magnetisation may then be found by taking suitable derivatives of f_s with respect to the temperature and the bulk magnetic field. Show that scaling implies that the exponents α_s and β_s characterising their critical behaviour are simply related to the analogous bulk exponents and ν.

7.3 What is the boundary scaling dimension of the magnetisation of the two-dimensional XY model in its low temperature phase (see Section 6.2), assuming (a) fixed, (b) free, boundary conditions on the spins? [Hint: use the method of images in the Gaussian model.]

7.4 What happens to the phase diagram of Figure 7.2 when the boundary is at its lower critical dimension (i.e. $d - 1 = 1$ and 2 respectively for systems with discrete or continuous symmetry)?

7.5 Consider a wedge-shaped geometry bounded by two planar surfaces meeting at an edge and subtending an angle α. For simplicity consider the Gaussian fixed point only, at which the correlation functions obey Laplace's equation with suitable boundary conditions, in this case $S = 0$ if the degrees of freedom near the boundary are undergoing the ordinary transition. By calculating the correlation function between a spin near the edge and one in the bulk, show that there are new *edge scaling dimensions* describing this behaviour, and that they depend continuously on the angle α.

7.6 Modify the ϵ-expansion of Section 7.3 to deal with the special transition. In this case, it is the field $h^{(1)}$ which is of interest, and the boundary condition is $\partial_z S = 0$.

8
Random systems

8.1 Quenched and annealed disorder

The systems discussed so far have been assumed to be homogeneous. Any real system will inevitably contain impurities. In most circumstances, one tries to eliminate them, but, under well-controlled conditions, it is also interesting to study their effect on critical behaviour. In general, one would expect any kind of random inhomogeneities to tend to disorder the system, and thus to lower the critical temperature. In fact, under certain circumstances, randomness may completely eliminate the ordered phase. Under other conditions, it is still possible for the system to order, but the universality class of the critical behaviour may be modified.

The first point to be made concerns the important distinction between annealed and quenched disorder. As a concrete example, suppose that we substitute some non-magnetic impurity atoms into a lattice of magnetic ions. The way we might do this is to mix some fraction of impurities into the melt, and let the system crystallise by cooling. If we allow this to happen very slowly, the impurities and the magnetic ions will remain in thermal equilibrium with each other, and the resulting distribution of impurities will be Gibbsian, governed by the final temperature and the various interactions between the different kinds of atom. Such a distribution of impurities is called *annealed*. If we study the thermodynamics of these very long time scales, we should do statistical mechanics in which we trace not only over the orientations of the spins of the magnetic ions, but also over the positions of the impurity atoms, in the partition function. However, in a solid the mobility of the impurities is so low that the time scales involved for them to reach positional thermal equilibrium are astronomical.

The more physical case corresponds to regarding the positions of the impurities as fixed, and tracing over only the magnetic degrees of freedom. This is the *quenched* case. As an example, consider an Ising ferromagnet where a fraction of the spins are replaced by vacancies. The reduced hamiltonian is of the form

$$\mathcal{H}(\{s\}, \{m\}) = -\tfrac{1}{2} \sum_{r,r'} J(r - r') m(r) m(r') s(r) s(r'), \qquad (8.1)$$

where $s(r) = \pm 1$ are the Ising spins, and the $m(r)$ are the quenched degrees of freedom, taking the values 0 or 1 according to whether or not there is an impurity at site r. The partition function is

$$Z(\{m\}) = \text{Tr}_s \, e^{-\mathcal{H}(\{s\}, \{m\})}. \qquad (8.2)$$

Note that we do not trace over the $m(r)$ – this would correspond to the annealed case – so that Z depends on the distribution of impurities. This makes life more difficult, since the system is no longer translationally invariant. However, in the thermodynamic limit, we are saved by the following result. All interesting thermodynamic quantities are given by suitable derivatives of the free energy $F(\{m\}) = -\ln Z(\{m\})$. For a large enough system we may imagine it subdivided into many subsystems which are still macroscopically large. Each subsystem will have a different distribution of impurities, which we may imagine as drawn out of some ensemble, with a probability distribution $P(\{m\})$. The total free energy is (apart from surface effects) the sum of the free energies of the subsystems. Thus, in the thermodynamic limit, the free energy per site of any system will, with probability one, be equal to the free energy averaged over the ensemble. This is called the *quenched average* free energy, defined by

$$\overline{F} = \text{Tr}_m \, P(\{m\}) F(\{m\}). \qquad (8.3)$$

As usual, the relative fluctuations away from this result for a particular system go like $V^{-1/2}$, where V is the volume of the system. This result is, in fact, only true if there are short range correlations between the impurities in the distribution function $P(\{m\})$, but this will usually be the case, since it is supposed to represent the high temperature distribution prior to the quench. For the purposes of studying the long distance physics of critical behaviour, it is sufficient to assume that the $m(r)$ at different sites are independent, so that we may take

$$P(\{m\}) = \prod_r \left((1-x)\delta_{m(r),0} + x\delta_{m(r),1}\right), \qquad (8.4)$$

where $1-x$ is the concentration of impurities.

The property of the free energy that its quenched average value represents its behaviour for a typical system is called *self-averaging*. Not all attributes of a random system have this property. For example the correlation function $\langle s(r_1)s(r_2)\rangle$, for fixed r_1 and r_2, is clearly very sensitive to whether there are impurities near r_1 or r_2, and is not always equal to its average value $\overline{\langle s(r_1)s(r_2)\rangle}$, even in the thermodynamic limit. However, the translationally averaged quantity $V^{-1}\sum_r \langle s(r_1+r)s(r_2+r)\rangle$ (whose Fourier transform enters the structure factor $S(q)$) is self-averaging.

The advantage of being able to average the free energy and similar quantities is that it restores the translational invariance of the problem.† However, things are still not simple, because we must average over the $m(r)$ in the free energy, rather than in the partition function, and this is not a straightforward operation in most cases. There is a way of dealing with this, through a subterfuge called the *replica method*. This is based on the formula

$$\ln Z = \lim_{n\to 0} \frac{Z^n - 1}{n}. \qquad (8.5)$$

The idea is then to average Z^n, which is often easier, and take the limit $n \to 0$ at the end of the calculation. The first step may be carried out, for positive integral n, by imagining n replicas of the system, in which the spin degrees of freedom $s_a(r)$ are labelled by an additional replica index a, with $1 \leq a \leq n$, but with each having the same configuration of impurities. Then

$$Z(\{m\})^n = \text{Tr}_{s_a} e^{-\sum_a \mathcal{H}(\{s_a\},\{m\})}, \qquad (8.6)$$

so that

$$\overline{Z^n} = \text{Tr}_m \text{Tr}_{s_a} P(\{m\})e^{-\sum_a \mathcal{H}(\{s_a\},\{m\})}. \qquad (8.7)$$

It is usually possible to carry out the trace over the $m(r)$ explicitly. This process then yields an effective short range hamiltonian

---

† This assumes that $P(\{m\})$ is translationally invariant. In real systems, there may be concentration gradients which are hard to eliminate.

for the replicated degrees of freedom, in which, however, the replicas are coupled together. This is the price we pay for restoring translational invariance to the problem. However, it allows all the standard techniques of critical behaviour in translationally invariant systems to be applied.

A word should be said concerning the validity of the replica approach. As long as it is used only as a book-keeping tool for generating weak disorder expansions (for example, in powers of $(1-x)$ in (8.7)), the coefficients in such an expansion are polynomials in n, and there is no difficulty in taking the limit $n \to 0$. In fact, such expansions may be derived independently of the replica method. However, problems may arise when non-perturbative techniques are used for finite n, which involve taking the thermodynamic limit $V \to \infty$ before the $n \to 0$ limit is taken. These two limits may not commute, and such methods may then give erroneous results. A classic example where this happens is in the spin glass problem. Even then, however, the approach may be made to work if the possibility of replica symmetry breaking is included.

8.2 The Harris criterion

One important question to address is whether the introduction of weak randomness changes the universality class of a transition – in renormalization group language, whether the impurities are relevant at the critical point of the pure system. Let us answer this question for a class of models in which the randomness couples to the local energy density. This includes the random site impurity model of equation (8.1), but also random bond models, where the exchange interaction $J(r - r')$ is random (as long as it is predominantly ferromagnetic). In the language of scaling operators near the critical fixed point of the pure system, the hamiltonian then has the form

$$\mathcal{H} = \mathcal{H}^* + \sum_r m(r)E(r), \qquad (8.8)$$

where $\mathcal{H}^*$ is the pure fixed point hamiltonian, $m(r)$ represents the random impurity field, and $E(r)$ is the local energy density. The quenched average of the replica partition function is given by

$$\overline{Z^n} = \text{Tr}_m \text{Tr}_{s_a} P(\{m\})e^{-\sum_a \mathcal{H}_a^* - \sum_a \sum_r m(r)E_a(r)} \qquad (8.9)$$

and the quenched average may be taken using the cumulant expansion

$$\mathrm{Tr}_m P(\{m\}) e^{-\sum_a \sum_r m(r) E_a(r)} = \qquad (8.10)$$
$$e^{-\overline{m} \sum_a \sum_r E_a(r) + \frac{1}{2} \sum_{ab} \sum_{rr'} (\overline{m(r)m(r')} - \overline{m}^2) E_a(r) E_b(r') + \cdots}.$$

The first term in the exponent on the right hand side adds a term proportional to the local energy density, which is therefore responsible for a shift in the value of T_c proportional to the density $(1-x)$ of impurities. Since the correlations between the impurities are short range, we may use the operator product expansion to write $E_a(r) E_b(r')$ as a sum of local terms. When $a = b$, the leading term is the energy density itself, which simply leads to a higher order shift in T_c. For $a \neq b$, however, this generates the important new term $\sum_{a \neq b} E_a(r) E_b(r)$.

To discover whether this is relevant, let us compute its scaling dimension via its 2-point correlation

$$\left\langle \sum_{a \neq b} E_a(r) E_b(r) \sum_{a' \neq b'} E_{a'}(r') E_{b'}(r') \right\rangle =$$
$$2n(n-1) \langle E_a(r) E_a(r') \rangle^2 \sim \frac{2n(n-1)}{|r - r'|^{4x_E}}. \quad (8.11)$$

The perturbing operator therefore has twice the scaling dimension of the energy operator at the pure fixed point, which, from (3.55), is $x_E = d - y_E = d - 1/\nu$. Thus the leading perturbation due to the randomness has renormalization group eigenvalue

$$y = d - 2x_E = 2/\nu - d. \qquad (8.12)$$

The higher terms in the cumulant expansion generate, through the operator product expansion, either the terms already discussed, or less relevant ones. Weak, locally correlated randomness which couples to the local energy is therefore irrelevant if $y < 0$, or

$$\boxed{d\nu > 2} \qquad (8.13)$$

where it should be emphasised that the exponent refers to the *pure* fixed point. (8.13) is known as the *Harris criterion*. When the pure system satisfies hyperscaling, (3.52) implies that the criterion is equivalent to $\alpha < 0$. For most systems in three dimensions, this is satisfied, and we therefore expect such weak randomness to have no effect on the critical exponents, although it will, of course, cause a shift in T_c. For the three-dimensional Ising model, however, $\alpha \approx 0.1$, and such randomness is relevant.

The Harris criterion does not, of course, say what will happen in cases when the randomness is relevant. The renormalization group derivation indicates that the cross-over exponent, defined in Section 4.2 is $\phi = 2 - d\nu$, but to what does the system cross over? Originally, it was thought that system might divide itself up into regions which undergo ordering at different temperatures. This would produce a smeared transition in which there would be no sharp singularities in thermodynamic quantities. However, both theoretical approaches, to be described below, and experiments, indicate that there is a sharp transition described by exponents differing from those of the pure system, and therefore described by a new, random, fixed point. The existence of such a fixed point does not, however, rule out the occurrence of other phenomena specific to a random system. For example, it has been proved that the free energy of a randomly diluted ferromagnet is singular, as a function of an applied external magnetic field, at all temperatures below the critical point of the *pure* system. These *Griffiths singularities* arise because there are arbitrarily large regions of the sample which contain no impurities at all, and therefore have a tendency to order close to the pure T_c. However, since such regions are extremely rare, these singularities are thought to be very weak, and therefore unobservable experimentally.

8.3 Perturbative approach to the random fixed point

We remarked above that the cross-over exponent for randomness coupling to the local energy is $2 - d\nu = \alpha$, assuming hyperscaling. For the Ising model, α is small is three dimensions, and it vanishes when $d = 2$, so that randomness is marginal there. This raises the possibility of applying the perturbative renormalization group of Section 5.2 in order to find the new random fixed point.

Let us therefore consider the Ising model in $2+\epsilon$ dimensions, for which the specific heat exponent of the pure model is α. We shall assume that this depends on ϵ in a smooth way, so that $\alpha = O(\epsilon)$ when ϵ is small. As explained above, the leading perturbation due to the randomness modifies the replica hamiltonian according to

$$\mathcal{H} = \mathcal{H}^* - \Delta \sum_r \sum_{a \neq b} E_a(r)E_b(r), \qquad (8.14)$$

where $\Delta \propto \overline{m^2} - \overline{m}^2 \propto x(1-x)$. It is important to note that, compared to the standard form chosen in (5.7), the sign of the coupling is *negative*. This is a common feature of replica hamiltonians. One might think that this change of sign means that the replica hamiltonian is not bounded from below, and therefore the partition function does not exist. However, the number of terms in the sum on the right hand side is $n(n-1)$, which itself changes sign below $n = 1$! We may choose to normalise the energy density so that $\langle E_a(r_1)E_b(r_2) \rangle \sim \delta_{ab}/r_{12}^{x_E}$. The renormalization group equation for Δ now follows from the operator product expansion of $\sum_{a \neq b} E_a E_b$ with $\sum_{c \neq d} E_c E_d$. Since the replicas are decoupled in $\mathcal{H}^*$, we may evaluate this using the operator product expansion of E_a with itself, the first few terms of which have the form

$$E_a \cdot E_b \sim \delta_{ab} + c\delta_{ab}E_a + \cdots, \qquad (8.15)$$

where c is a coefficient whose value is fixed and universal, once the normalisation of E is fixed as we have done. The term in the operator product expansion which we need therefore comes either from contracting just one pair, for example E_a with E_c, using the first term in (8.15), or from contracting two pairs and using the second term. The coefficient is just a matter of counting, and we find

$$\left(\sum_{a \neq b} E_a E_b \right) \cdot \left(\sum_{c \neq d} E_c E_d \right) \sim \left(4(n-2)+2c^2 \right) \sum_{a \neq b} E_a E_b + \cdots, \quad (8.16)$$

where the neglected terms contribute, for example, to the renormalization of t to $O(\Delta^2)$ and are therefore of higher order in ϵ. The renormalization group equation for Δ is thus

$$d\Delta/d\ell = \alpha\Delta + \left(4(n-2) + 2c^2 \right) \Delta^2 + O(\Delta^3, \epsilon\Delta^2). \qquad (8.17)$$

The plus sign in front of the second term, as compared with the standard form (5.14), reflects the different sign of the perturbation. It turns out that c vanishes when $d = 2$. This is a consequence of the self-duality of the Ising model in two dimensions. This duality is very similar to that discussed between the XY and SOS models in Section 6.3, except that the Ising model is its own dual.

The duality transformation maps high temperatures into low temperatures, thus reversing the sign of the reduced temperature t. Since this couples linearly to E, duality also reverses the sign of E. This in turn implies that, at the critical point which is self-dual, E cannot appear in the operator product expansion $E \cdot E$, since the two terms transform differently. If we once again assume that c is a smooth function of ϵ, we may therefore take it be $O(\epsilon)$ when this is small. As a result, to first order in ϵ, the term involving c^2 may be neglected. Note that the coefficient of the $O(\Delta^2)$ term in the beta function depends analytically on n, and there is no problem in taking the limit $n \to 0$. Having done this, we see that there is a non-trivial fixed point at $\Delta^* = \alpha/8 + O(\epsilon^2)$, which we interpret as the new random fixed point.

To find the change in the thermal exponent, we need the operator product expansion with the energy operator

$$\sum_{a \neq b} E_a E_b \cdot \sum_c E_c \sim 2(n-1) \sum_c E_c + \cdots, \qquad (8.18)$$

so that the renormalization group equation for t becomes

$$dt/d\ell = y_t^p t + 4(n-1)\Delta t + \cdots, \qquad (8.19)$$

where $y_t^p = d/(2-\alpha) \approx 1 + \frac{\epsilon}{2} + \frac{\alpha}{2}$ is the thermal eigenvalue at the pure fixed point. Setting $n = 0$ and $\Delta = \Delta^*$ in (8.19), the first order correction to this at the random fixed point is $-\frac{\alpha}{2}$. Thus, to first order in ϵ, the random exponents are $\nu' = 1 - \frac{\epsilon}{2} + O(\epsilon^2)$ and $\alpha' = 0 + O(\epsilon^2)$. In fact, there is a rigorous theorem that $d\nu' \geq 2$ at any random fixed point, implying that $\alpha' \leq 0$. There is no $O(\Delta)$ renormalization of the magnetic field h, since the operator product expansion of $E_a E_b$ with the magnetisation s_c cannot result in a term linear in s_c. Thus y_h is unchanged from its pure value, to first order in ϵ. This implies, by the scaling laws of Section 3.5, that $\beta'/\beta = \nu'/\nu = 1 + \frac{\alpha}{2}$, to this order. Clearly it is unrealistic to apply the above results to the case of three dimensions without examination of the higher order terms. However, the above calculation does simply illustrate the mechanism whereby the random fixed point may arise.

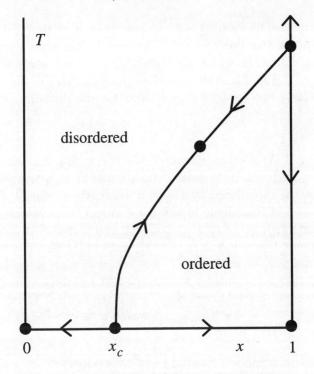

Figure 8.1. Phase diagram of a dilute ferromagnet.

8.4 Percolation

So far, in this chapter, we have considered only the case of weak randomness coupling to the local energy density. If the density of impurities is high, however, we may envisage a situation where there are only finite clusters of magnetic ions, so that a typical magnetic ion is coupled to only a finite number of others. In such a situation the clusters will behave independently, leading to para-magnetism. The critical concentration x_c below which this happens is called the *percolation threshold*. The phase diagram in the (x, T) plane of a typical dilute ferromagnet is shown in Figure 8.1. The critical temperature approaches zero as $x \rightarrow x_c+$.

Percolation itself does not involve the temperature, and is a purely geometrical phenomenon. However, it turns out that it has many features completely analogous to those of a continuous ther-mal phase transition, with the concentration x playing the role of temperature. The clearest way to see this connection is through a mapping to the *Potts model*. This model is a generalisation of the Ising model, in which the spins $s(r)$ at each site may be in any one of Q possible states, labelled, for example, by the integers $(1, 2, \ldots, Q)$. The reduced hamiltonian for the Potts model is

$$\mathcal{H} = -\sum_{r,r'} J(r - r')\delta_{s(r),s(r')}, \qquad (8.20)$$

where $J > 0$. At low temperatures, the energy is minimised by the spins $s(r)$ all being in the same state, while at high temperatures the system is disordered by thermal fluctuations. For $Q = 2$, the Potts model is equivalent to the Ising model, since we may write in that case $\delta_{s(r),s(r')} = \frac{1}{2}(s(r)s(r')+1)$, and therefore it undergoes a continuous transition at a critical value of J. The Potts model may be defined for arbitrary Q, not necessarily an integer, in the following manner. Consider for definiteness an interaction J which connects nearest neighbours only on a given lattice. Then we may write $e^{J\delta_{s(r),s(r')}} = 1 + x\delta_{s(r),s(r')}$, where $x \equiv e^J - 1$, so that the partition function becomes

$$Z(Q, x) = \mathrm{Tr}_s \prod_{\mathrm{n.n}} \left(1 + x\delta_{s(r),s(r')}\right). \qquad (8.21)$$

This may be expanded in powers of x as follows. If there are N nearest neighbour bonds on the lattice altogether, the expansion of the right hand side leads to 2^N terms. Each may be represented by a graph drawn on the lattice, in which a given bond is included or not, according as to whether the second or the first term is chosen in (8.21) for that particular bond. Each included bond will carry a factor of x as well as a delta function enforcing the equality of the spins on the sites which it connects. A typical configuration on the square lattice is shown in Figure 8.2. In general, the connected bonds will form clusters, and, within each cluster, the delta functions will force the spins at each vertex to be the same. Therefore, the trace over the spins $s(r)$ will lead to a factor of Q for each cluster, and the partition function may be written as

$$Z(Q, x) = (1 + x)^{-N} \sum_{\mathcal{C}} Q^{N_c} x^{N_b}, \qquad (8.22)$$

where the sum is over distinct cluster configurations, and N_c is the total number of clusters (note that single isolated sites count

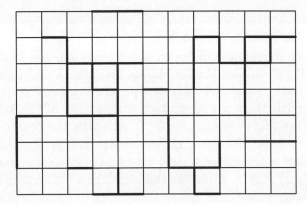

Figure 8.2. Diagram representing a typical term in the expansion of (8.21).

as clusters by this definition). N_b is the total number of bonds in the configuration. The factor of $(1 + x)^{-N}$ is included for later convenience. Since it depends analytically on x, it does not affect the critical behaviour. For a finite lattice, (8.22) is a polynomial in Q and therefore may be defined for non-integer values.

This maps onto the problem of *bond percolation*. In this version of the percolation problem, the bonds are independently distributed, each being present with probability p (and absent with probability $1-p$). The weight for a given configuration is therefore $p^{N_b}(1 - p)^{N-N_b}$. The two problems are evidently equivalent if we choose $x = p/(1 - p)$, and $Q = 1$. However, exactly at $Q = 1$, the partition function is trivial, $Z = 1$, since the probabilities of all configurations must sum up to unity. In order to obtain non-trivial information about percolation, we must consider the *limit* $Q \to 1$. To see this, suppose that $Q = 1 + \epsilon$, where $\epsilon \ll 1$. To first order in ϵ,

$$Z = 1 + \epsilon \sum_{\mathcal{C}} \sum_{\text{clusters in } \mathcal{C}} P(\mathcal{C}), \qquad (8.23)$$

where $\mathcal{C}$ denotes a configuration of clusters, and $P(\mathcal{C}) = p^{N_b}(1 - p)^{N-N_b}$ is the probability of its occurrence. The sum over the clusters in $\mathcal{C}$ gives a factor of $N_c(\mathcal{C})$, the number of clusters in the configuration. Therefore $\partial \ln Z / \partial Q$, evaluated at $Q = 1$, gives the mean number of clusters $\langle N_c \rangle$. (This also follows directly from (8.22).) The necessity of taking such a limit is clearly reminiscent of the $n \to 0$ replica method for evaluating the quenched free energy, described in Section 8.1.

Further connections between percolation and the Potts model emerge if we consider the correlation function of the order parameter. In fact, since there are Q equivalent ground states in the Potts model, there are Q distinct components of the local order parameters $M_a(r) = \delta_{s(r),a} - Q^{-1}$, satisfying the constraint that $\sum_a M_a(r) = 0$. The term Q^{-1} is subtracted so that $\langle M_a \rangle = 0$ in the high temperature phase, where all states are equally likely and $\langle \delta_{s(r),a} \rangle = Q^{-1}$. Consider the correlation function in the high temperature phase

$$G_{aa}(r_1, r_2) = \left\langle \left(\delta_{s(r_1),a} - Q^{-1} \right) \left(\delta_{s(r_2),a} - Q^{-1} \right) \right\rangle . \qquad (8.24)$$

If we evaluate this in the cluster expansion described above, the contribution from a given configuration will vanish if r_1 and r_2 lie in different clusters, because then the independent sum over $s(r_1)$ and $s(r_2)$ will give zero. But if they happen to lie in the same cluster, the values of the two spins are constrained to be equal, and we obtain a contribution $\sum_s (\delta_{s,a} - Q^{-1})^2 = (Q-1)/Q$. If we therefore define the correlation function of the percolation problem as the probability $G(r_1, r_2)$ that the sites r_1 and r_2 are in the same cluster, we have the connection

$$G(r_1, r_2) = \lim_{Q \to 1} (\partial/\partial Q) G_{aa}(r_1, r_2). \qquad (8.25)$$

The advantage of this mapping is that we may assume that the Q-state Potts model has a continuous transition,† similar to that of the Ising model (but with different exponents), and explore all the scaling consequences of a renormalization group fixed point. The analogue of the reduced temperature variable is played by $p_c - p$, where p_c is the percolation threshold. This analogy implies, for example, that for $p < p_c$, G decays asymptotically exponentially with increasing separation $|r_1 - r_2|$, with a correlation length ξ which diverges as $p \to p_c$ like $(p_c - p)^{-\nu}$. The average number of sites in a given cluster is $\sum_{r'} G(r, r')$, which, by (8.25), is given by $\lim_{Q \to 1} (\partial/\partial Q) \chi(Q, p)$, where χ is the susceptibility of the Potts model. This diverges as $A(Q)(p_c - p)^{-\gamma(Q)}$, where, in general, both the amplitude and the exponent depend continuously on Q.

---

† In fact, the model has a first order transition for sufficiently large Q, but not for $Q \leq 2$.

However, for the limit to make sense, $A(1) = 0$, so that, after differentiating with respect to Q, we see that the mean number of sites in a cluster diverges as $A'(1)(p_c - p)^{-\gamma(1)}$. In general, the exponents of percolation are those of the Q-state Potts model *at* $Q = 1$, not their derivatives with respect to Q.

One may ask not only about the mean number of sites in a cluster, but, more generally, about the *cluster size distribution*. Let $n_s(p)$ be the total number of clusters containing exactly s sites, divided by the total number of sites on the lattice. Then, for example, $\sum_s n_s$ is the total number of clusters per unit volume. From (8.22) this is given by the derivative of the free energy of the Potts model with respect to Q, evaluated at $Q = 1$, and it therefore has a singular part proportional to $(p_c - p)^{2-\alpha}$. Similarly, $\sum_s s n_s$ just gives the proportion of sites in any (finite) cluster, and $\sum_s s^2 n_s$ gives the mean cluster size we calculated above. In general, the moment $\sum_s s^k n_s$ is given by a derivative with respect to Q of the k-point correlation function $\langle M_a(r_1) \dots M_a(r_k) \rangle$, summed over $k-1$ of its points. According to the transformation law (3.59) this will scale like $\xi^{(k-1)d - k x_M}$, where $\xi \propto (p_c - p)^{-\nu}$, and x_M is the scaling dimension of M, related to the standard exponents in the usual way (see Chapter 3). This result for the moments of $n_s(p)$ is consistent with the scaling hypothesis for the cluster size distribution

$$n_s(p) \sim s^{-\tau} f\left((p_c - p)s^\sigma\right). \tag{8.26}$$

This is supposed to be valid as $s \to \infty$ and $p \to p_c$. The asymptotic behaviour of the moments may be then be simply estimated by replacing the sums over s by integrals. Comparison with the above results then shows that σ and τ are related to to the standard exponents by

$$1/\sigma = (d - x_M)\nu = \beta + \gamma \tag{8.27}$$
$$(\tau - 1)/\sigma = d\nu = 2 - \alpha. \tag{8.28}$$

What of the low temperature ordered phase? We expect the Q-fold permutation symmetry of the Potts model to be spontaneously broken. A simple way of defining the magnetisation is to consider initially a finite geometry, and to fix the boundary spins to be in a given state, say a. If we now measure $\langle M_a(r) \rangle$ for an internal site r far from the boundary, in the thermodynamic limit this will be zero in the high temperature state and non-zero below T_c, vanishing as $(T_c - T)^\beta$. Thus, suitable boundary conditions are equally as effective in selecting a unique low temperature state as a uniform external field. From the point of view of percolation, fixing all the boundary spins into the same state means that they are in the same cluster and therefore all connected. If b is any boundary site, since $M_a(b) = 1 - Q^{-1}$, we may write $\langle M_a(r) \rangle$ equally well as $Q(Q-1)^{-1} \langle M_a(r) M_a(b) \rangle$, where b is any site on the boundary. In the limit $Q \to 1$ this is, by (8.25), the probability $G(r, b)$ that r is connected to b. We conclude that the analogue of the spontaneous magnetisation for percolation is the probability that a site deep inside the system is connected to the boundary, in the thermodynamic limit, and that this quantity will vanish as $(p - p_c)^{\beta(Q=1)}$. For such a site to be connected to the boundary there must exist a cluster in which both of them lie, and this must persist in the thermodynamic limit. Thus, for $p > p_c$, there must exist at least one 'infinite cluster' which contains a finite fraction $\langle M \rangle$ of the sites. In fact, there is only one such cluster, and $\langle M \rangle$ may thus be interpreted as the probability that a given site belongs to it.

All of this relies on the assumption of the existence of a critical fixed point with similar properties to that of the Ising model. This assumption may be investigated by studying the model in the vicinity of its upper critical dimension. To do this we must formulate an appropriate continuous spin version of the Potts model. Since only the qualitative form of this is needed, the result may in fact be written down on the basis of symmetry arguments alone. Introduce continuous spin variables $S_a(r)$, with $1 \leq a \leq Q$, whose expectation values will be the order parameters M_a. These must satisfy the constraint $\sum_a S_a = 0$. The most relevant terms in the continuous spin hamiltonian will be those with the lowest powers of S_a and the fewest derivatives. In addition, terms should be classified according to how they transform under the permutation group of Q objects. The allowed invariant quadratic terms are of the form

$$\mathcal{H}_0 = \int \left[\sum_a (\nabla S_a)^2 + t \sum_a S_a^2 \right] d^d r, \qquad (8.29)$$

just as in the $O(n)$ model of Section 5.7. Note that a term such as $\sum_{a \neq b} S_a S_b$, while allowed by the symmetry, may be rewritten, using the constraint $\sum_a S_a = 0$, in the above form. However, at the level of cubic invariants, the Potts model differs from the $O(n)$ model. Various terms are possible, for example, $\sum_{a \neq b} S_a^2 S_b$, $\sum_{a \neq b \neq c} S_a S_b S_c$, and so on, but these may all be written, using the constraint, in terms of $\sum_a S_a^3$. The most relevant interaction term to be added to (8.29) is therefore of the form $u_3 \int \sum_a S_a^3 d^d r$. The existence of such a cubic invariant implies that, in mean field theory, the Potts model has a first order transition, at least for $Q > 2$. However, this argument is not directly applicable to the case of percolation, since it is based on minimising a free energy which has no direct physical meaning for that problem. In fact, the percolation problem is believed to exhibit a continuous transition in all dimensions.

In writing down $\mathcal{H}$ the explicit factors of the lattice cut off a have not been included, but they may easily be restored by dimensional analysis. If we choose to normalise S_a so that the coefficient of the $(\nabla S)^2$ term is unity, then S_a has dimension $(\text{length})^{1-d/2}$, so that, as usual, t has dimension $(\text{length})^{-2}$, and u_3 dimension $(\text{length})^{d/2-3}$. The upper critical dimension for the Potts model, and therefore for percolation, is therefore $d = 6$ rather than $d = 4$. The thermal eigenvalue corresponding to the scaling field t, which is proportional to $(p_c - p)$, is $y_t = 2$ at $d = 6$, corresponding to $\nu = 1/2$. The magnetic eigenvalue at the Gaussian fixed point, corresponding to an external symmetry breaking field $h S_a$, is $y_h = d/2 + 1$, as in Section 5.4, but, in $d = 6$, this becomes $y_h = 4$. As usual, the mean field exponents, which are valid for $d \geq 6$, are those of the Gaussian fixed point at $d = d_c$. This leads to the values $\alpha = -1$, $\beta = 1$, $\gamma = 1$, $\nu = \frac{1}{2}$ and $\eta = 0$. Although the expansion in powers of $\epsilon = 6 - d$ below the upper critical dimension is clearly not very reliable for the numerical values for exponents in two or three dimensions, the very existence of a nontrivial fixed point implies that they should satisfy the usual scaling laws described in Section 5.5.

Cross-over to finite temperature

As described above, percolation is a purely geometrical phenomenon with direct relevance for dilute ferromagnets only ex-

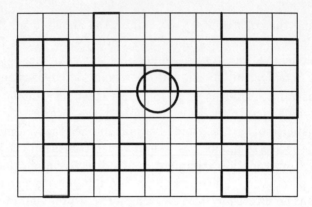

Figure 8.3. Pivotal bond in a percolation cluster.

actly at $T = 0$.† At non-zero temperature, the fluctuations of the magnetic degrees of freedom must also be taken into account, and the question arises as to whether these effects are relevant in the renormalization group sense.

At low temperatures, only the lowest energy excitations are important. Near the percolation threshold, these correspond to flipping all the spins in the part of a cluster which is connected to the rest of the cluster by only a single bond, as shown in Figure 8.3. The Boltzmann weight for such excitation is e^{-2K}, where K is the reduced nearest neighbour exchange coupling (so that $K \propto 1/T$). The occurrence of such 'pivotal' bonds is related to the expectation value of the energy operator $E(r, r') = \delta_{s(r), s(r')}$ of the Potts model (where r and r' are nearest neighbour sites) as follows. Consider inserting $E(r, r')$ into a given configuration of clusters. If r and r' are already in the same cluster, $E(r, r') = 1$ and the bond, if added to the configuration, is not pivotal. However, if the sites are in different clusters, the bond will be pivotal, and inserting $E(r, r')$ will give one less factor of Q when the trace over the spins is performed. So we may say that $E(r, r') = Q^{-1}$ in this case. The same is true of multiple insertions of E: they are significant only when they are made on pivotal bonds. We may therefore write the low-temperature expansion of the partition

---

† In stating this we are of course considering only the classical description of a magnet. At sufficiently low temperatures, quantum effects must also be taken into consideration.

function, in powers of e^{-2K}, as a sum over correlation functions of the energy density of the Potts model. The scaling dimension associated with such a perturbation is therefore the same as that of the energy density itself, which is given by $x_E = d - y_t$. The renormalization group eigenvalue associated with e^{-2K} is therefore just the 'thermal' eigenvalue y_t of the Potts model at $Q = 1$. The effects of non-zero temperature are therefore *relevant*. For p slightly greater than p_c, the critical behaviour at the ferromagnetic transition will exhibit a cross-over from percolative exponents to another universality class. The simplest assumption is that the renormalization group flows in this case end at the random fixed point, which will be distinct from that of the undiluted system if the Harris criterion is not satisfied. The renormalization group flows in this case, projected onto the (x, T) plane, are shown in Figure 8.1. The cross-over exponent for the finite temperature cross-over is simply $\phi = y_t \nu = 1$, a result which has been verified by explicit renormalization group calculations. This implies that the shape of the phase boundary in the (x, T) plane near the percolation threshold has the form $T \propto (\ln(p - p_c))^{-1}$.

8.5 Random fields

So far, we have discussed the effects of randomness which couples to the local energy density, as exhibited, for example, by a dilute ferromagnet. A much more severe type of disorder occurs if the randomness couples to the local order parameter, for example, to the local magnetisation in a ferromagnet. A simple example is given by the Ising model in a random field, with hamiltonian

$$\mathcal{H} = -\sum_{r,r'} J(r, r')s(r)s(r') - \sum_r h(r)s(r), \qquad (8.30)$$

where, as usual, $s(r) = \pm 1$. The $h(r)$ are quenched random variables, such that their mean $\overline{h(r)} = 0$, and are usually taken to have only short range correlations. The random field Ising model may be experimentally realised using a binary fluid in a gel which has a preferred affinity for one of the components, or, through a clever mapping, as a dilute antiferromagnet in a uniform external field.

For weak randomness, the replica method and the cumulant expansion may be used to calculate the cross-over exponent, as in Section 8.2. Instead of coupling to the product of energy densities $\sum_{a \neq b} E_a E_b$, the second cumulant $\overline{h^2}$ now couples to $\sum_{a \neq b} S_a S_b$, which has scaling dimension (at the pure fixed point) $2x_h = 2(d - y_h)$, using the scaling laws of Section 3.5. Thus its renormalization group eigenvalue is $y = 2y_h - d$, and the cross-over exponent is $y/y_t = \gamma$.† This is always positive, so that random fields are always relevant.

At high temperatures, the spins will tend to follow the direction of the local random field. One might ask whether, in fact, they may order at any finite, or indeed, at zero temperature. This is determined by the Imry–Ma criterion, which is a generalisation of the domain wall arguments given in Section 6.1 for the lower critical dimension. Consider a random field Ising model in d dimensions, and assume that it is ordered, with the majority of the spins $+1$ at zero temperature. Suppose we introduce a region R, of linear size L, within which the spins are reversed. As before, the energy will increase, by an amount of order JL^{d-1}, due to the introduction of this domain wall. It will also change by an amount $\sum_{r \in R} h(r)$ due to the interaction with the random field. The quantity $\sum_{r \in R} h(r)$ is, for large domains, a normally distributed random variable with zero mean, but a typical value $\pm(\overline{h^2}L^d)^{1/2}$. If we are judicious in choosing the region R, we may ensure that this represents a lowering of the energy. Comparing the L-dependence of these two contributions, we see that, for $d < 2$, the introduction of such a large domain can always lower the energy, even for very weak random fields, while, for $d > 2$, the ordered state would appear to be stable. The Imry–Ma argument thus predicts that the lower critical dimension of the random field Ising model is $d_l = 2$.

This suggests that the critical random field fixed point should be accessible perturbatively in $2 + \epsilon$ dimensions, just as for the $O(n)$ model discussed in Section 6.5. The analysis is rather different however, since the random field fixed point occurs at *zero temperature*, even when $d > 2$. To see this consider the renor-

---

† For dilute Ising antiferromagnets, the relevant fixed point about which to perturb is the random one described in Section 8.3. In this case, the cross-over exponent is not exactly γ, since the replicas are already coupled by the bond randomness.

malization group equations for the random field strength h_R, the exchange coupling J, and a possible uniform magnetic field h. The critical point occurs when $h = 0$, and can depend only on the dimensionless ratio $w \equiv h_R/J$. The same arguments which led to the Imry–Ma result determine the lowest order terms in the renormalization group equations

$$dh_R/d\ell = (d/2)h_R \qquad (8.31)$$
$$dJ/d\ell = J((d-1) - Aw^2 + \cdots). \qquad (8.32)$$

The $O((h_R/J)^2)$ term in (8.32) arises from a renormalization of the energy per unit area of a domain wall, reflecting the local distortion of the wall as it adjusts itself so as best to suit the local configuration of the random field. The coefficient A is model-dependent and its value is not important, except that it is positive, reflecting a reduction in the domain wall energy. It may be argued that there are no similar $O(w^2)$ corrections to (8.31), nor to the renormalization group equation for the uniform field

$$dh/d\ell = d\,h, \qquad (8.33)$$

which follows since this is a discontinuity fixed point (see Section 4.1). The renormalization group equation for w itself therefore has the form

$$dw/d\ell = -\tfrac{\epsilon}{2}w + Aw^3 + \cdots, \qquad (8.34)$$

showing the existence of a perturbative fixed point $w^{*2} = O(\epsilon)$, as expected. Linearising about this gives the eigenvalue $\epsilon + O(\epsilon^2)$. This may be identified with the exponent $1/\nu$ in the usual way, where ν controls the divergence of the correlation length $\xi \propto |w - w^*|^{-\nu}$ at zero temperature. The effects of finite temperature T may be discussed by noting that it appears in the dimensionless combination T/J. Therefore the renormalization group equation for T follows from (8.32):

$$d(T/J)/d\ell = -(T/J)((d-1) - Aw^2 + \cdots) = -\theta(T/J) \qquad (8.35)$$

where $\theta = 1 + \tfrac{\epsilon}{2} + \cdots$ at the fixed point. The temperature is thus *irrelevant* at the random field fixed point, reflecting the fact that the disorder due to the random field overwhelms the thermal fluctuations.

The phase diagram in the (T, w) plane of the random field Ising model for $d > 2$ is therefore expected to be as shown in Figure 8.4.

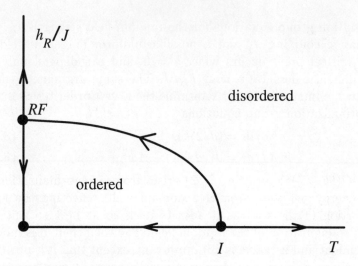

Figure 8.4. Phase diagram and projected RG flows for the random field Ising model for $d > 2$.

The significance of this is that the zero-temperature fixed point controls the whole of the critical line $T = T_c(w)$ for $w > 0$, so that, for example, the exponent ν calculated above also determines the divergence of the correlation length as $T \to T_c(w)$ at fixed w.

The fact that the critical fixed point occurs at zero temperature means that the scaling equation for the reduced free energy derived in Section 3.4 is not appropriate. Instead, we should discuss the dimensionless energy density $e(w, h/J) \equiv J^{-1}E(w, h/J)$. Under a rescaling by a factor b, E scales with its usual factor b^{-d} (see (3.25), but there is an additional factor of b^θ coming from the rescaling of J. This has the consequence that d is replaced by $d-\theta$ in many of the usual scaling laws. For example, α, defined by $e \propto |w - w^*|^{2-\alpha}$, is related to ν by the modified hyperscaling law (cf. (3.52))

$$2 - \alpha = (d - \theta)\nu. \tag{8.36}$$

This violation of simple hyperscaling may be traced to the existence of the dangerous irrelevant scaling variable T, see p.49.

Once again, the study of this model near its upper critical dimension gives further information. In this case, it is more appropriate to use a continuum spin version. After averaging over the random field, this corresponds to the replica hamiltonian

$$\mathcal{H} = \int \left(\sum_a [(\nabla S_a)^2 + tS_a^2 + uS_a^4] - \Delta \sum_{ab} S_a S_b \right) d^d r, \quad (8.37)$$

where $\Delta \propto \overline{h^2}$. It turns out to be more convenient to include the contributions with $a = b$ in the last term. This modification merely shifts the temperature variable t. Let us first consider the Gaussian model, with $t = u = 0$. Dimensional analysis implies that Δ is relevant, with renormalization group eigenvalue 2, at the pure Gaussian fixed point with $\Delta = 0$. In the absence of any interaction terms, it will keep on flowing all the way to infinity. However, it does not make sense to set it equal to infinity in (8.37), and we must therefore do the calculations keeping it finite. The correlation functions in this Gaussian model are found, by the rules of the Appendix, by Fourier transforming the quadratic terms and then inverting the matrix in replica space

$$\mathbf{G}^{-1} = k^2 \mathbf{1} - \Delta \mathbf{M}, \quad (8.38)$$

where $\mathbf{M}$ is a matrix all of whose elements are unity. This inversion is easily performed by expanding in powers of Δ and noting that $\mathbf{M}^2 = n\mathbf{M}$. The result is

$$\mathbf{G} = \frac{1}{k^2} + \frac{\Delta \mathbf{M}}{k^2(k^2 - n\Delta)}. \quad (8.39)$$

The diagonal and off-diagonal elements of this matrix have different physical interpretations. A scattering experiment would measure the Fourier transform of the quenched average of the correlation function $\overline{\langle S(r_1)S(r_2)\rangle}$, which is equal to G_{aa}. This contains, in the limit $n \to 0$, both $1/k^2$ and $1/k^4$ pieces. The latter leads, when $t \neq 0$, to a (Lorentzian)^2 line shape which is characteristic of random field systems. On the other hand, the susceptibility is given by the sum over r_1 and r_2 of $\overline{\langle S(r_1)S(r_2)\rangle} - \overline{\langle S(r_1)\rangle\langle S(r_2)\rangle}$, which equals $G_{aa} - G_{ab}$, with $a \neq b$, and is thus given by only the first term in (8.39). This shows that the results for exponents at the Gaussian fixed point, namely $\gamma = 1$, $\nu = \frac{1}{2}$ and $\eta = 0$, are the same as for the pure system. On the other hand, mean field theory also gives the usual result $\alpha = 0$, since the last term in (8.37) is of order n^2, and therefore does not contribute to the quenched free energy $\lim_{n \to 0} n^{-1} \ln Z$.

However, the fact that the Fourier transform of the correlation function behaves at $n = 0$ as $1/k^4$ rather than the more usual $1/k^2$ has a dramatic effect on the fluctuations. For example, the calculation of the upper critical dimension through the Ginzburg criterion, described in Section 2.4, is modified by two extra powers of k in the denominator of (2.20), with the result that the upper critical dimension is *six*, rather than four. The usual argument, based on the relevance of u, fails because of the existence of the additional dimensionful parameter Δ, which cannot be scaled out of the problem. In fact, this is the way in which the dangerous irrelevant variable shows up in this formulation of the problem. Again, one consequence of this is that the usual hyperscaling relation $\alpha = 2 - d\nu$ is not satisfied near $d = 6$. Instead, it is replaced by a modified relation (8.36), with $\theta = 2$ at $d = 6$.

The ϵ-expansion below $d = 6$ may be carried out in a similar manner to that described in Section 5.5 for the pure case. The main difference is that, whenever two S_a are contracted, a factor Δ enters, according to (8.39). Thus the effective expansion parameter in perturbation theory is not u but the product $u\Delta$. It is this combination which is marginal at $d = 6$, and which goes to a non-trivial fixed point value in $6-\epsilon$ dimensions, even though u and Δ are each scaling towards zero and infinity, respectively. We may make contact with the arguments of the previous section if we reinstate the temperature T in the parameters of the reduced hamiltonian (8.37): u, after rescaling $S^2 \to TS^2$, will then contain a factor T, while Δ, being proportional to $\overline{h^2}$, will go like $1/T$. Thus the random field fixed point occurs at $T = 0$, as expected.

However, according to the ϵ-expansion, the exponent θ remains fixed at the value two, to all orders in perturbation theory. A related, and remarkable, fact is that the exponents of the random field fixed point in d dimensions are the same, order by order in ϵ, as those of the pure fixed point in $d-2$ dimensions. If this result were valid beyond perturbation theory, it would imply that the lower critical dimension of the random field Ising model is larger by two than that of the pure model, that is, equal to three. This, of course, contradicts the Imry–Ma argument discussed above (and more rigorous versions thereof), and the result $\theta = 1 + \frac{\epsilon}{2} + \cdots$ found there. Therefore the simple ϵ-expansion for the random field prob-

lem must break down. How it does so is not completely understood, although recent results suggest that it is connected with
the spontaneous breaking of replica symmetry – a phenomenon
which is known to occur for the spin glass problem.

Exercises

8.1 Consider the one-dimensional Ising model, with a concentration $1 - x$ of non-magnetic impurities. Note that, in one
dimension, these impurities completely sever the magnetic
correlations. Starting from the result found in Ex. 4.2 for the
susceptibility of a finite length chain, calculate the quenched
average susceptibility of the system as a function of the concentration x and the correlation length ξ of the pure system.

8.2 Generalise the Harris criterion to the case when the impurities are correlated, with a density-density correlation function which decays like $r_{12}^{-d-\sigma}$. Similarly, adapt it to the case
of uncorrelated random impurities in a *quantum* system at
low temperatures, using the mapping to a $(d+1)$-dimensional
classical system discussed in Section 4.5.

8.3 According the arguments on p.151, dilution is marginally irrelevant for the two-dimensional Ising model. Generalise the
argument of Ex. 3.6 to show that in this case the specific heat
diverges like $\ln \ln |t|$.

8.4 In the continuous spin version of the Ising model, the coupling to quenched randomness corresponds, in the replica formalism, to a term $-\Delta \sum_{ab} S_a^2 S_b^2$. Show that the resultant
hamiltonian is identical to the $n \to 0$ limit of the $O(n)$ model
with cubic symmetry breaking, discussed in Section 5.6. Unfortunately, these first-order renormalization group equations
have no non-trivial random fixed point, and it is necessary
to go to higher orders to find one. Show that this difficulty
does not arise if instead one considers the $O(m)$ model with
random impurities, with $m \neq 1$.

8.5 Consider the two-dimensional Gaussian model of p.114
with an additional term in the hamiltonian proportional to
$h_p \int \cos(p\theta - \phi(r)) d^2 r$, where $\phi(r)$ is a quenched random
phase, uniformly distributed over the interval $[0, 2\pi)$. [Such

a term models the effect of a random substrate on a two-dimensional incommensurate solid.] Using the replica trick, discuss for what values of K this perturbation is relevant, and calculate its cross-over exponent.

8.6 Suppose a percolation problem is formulated in the interior of a rectangular region. Show that the probability that a given edge of the rectangle is connected to the opposite edge is related to a difference of partition functions of the Potts model with different boundary conditions. [Hint: suppose that all the sites on a given edge are connected with probability one to an extra site.] Hence write down a finite-size scaling form for the required probability.

8.7 Using the mapping to the Potts model, define as many universal amplitude combinations for percolation as you can.

9

Polymer statistics

Linear polymers may be thought of as very long flexible chains made up of single units called monomers. When placed in a solvent at low dilutions, they may exhibit several different types of morphology. If the interactions between different parts of the chain are primarily repulsive, they tend to be in extended configurations with a large entropy. If, however, the forces are sufficiently attractive, the chains collapse into compact objects with little entropy. The collapse transition between these two states occurs at the theta point, where the energy of attraction balances the entropy difference between the two states. This turns out to be a continuous phase transition, to be described later in Section 9.5. However, even in the swollen, entropy dominated, phase, it turns out that the statistics of very long chains are governed by non-trivial critical exponents. Like the percolation problem, this is a purely geometrical phenomenon, yet, through a mapping to a magnetic system, all the standard results of the renormalization group and scaling may be applied. Before describing this, however, it is important to recall some of the simpler approaches to the problem.

9.1 Random walk model

If the problem of a long polymer chain is equivalent to some kind of critical behaviour, we would expect universality to hold, and some of the important quantities to be independent of the microscopic details. This means that we may forget all about polymer chemistry, and regard the monomers as rigid links of length a, like a bicycle chain. In the simple random walk model, one makes one further assumption: that all interactions between different parts of the polymer may be neglected. The most important of these are the steric repulsions which dictate that two pieces of the polymer

169

cannot be in the same place at the same time. This approximation will be valid if the probability of this happening is small, a
condition which may be examined *a posteriori*. Suppose each link
is oriented along a vector ρ_i, where $i = 1, 2, \ldots, N$ and $|\rho_i| = a$. In
the random walk model, these are independent random variables,
each with a uniform distribution of their orientations. A simple
quantity to calculate is the mean square distance between the two
ends:

$$\langle R^2 \rangle = \left\langle \left(\sum_i \rho_i \right)^2 \right\rangle = \sum_i \langle \rho_i{}^2 \rangle + \sum_{i \neq j} \langle \rho_i \cdot \rho_j \rangle = N a^2, \qquad (9.1)$$

since the second term vanishes. Thus the rms end-to-end distance
behaves like N^ν, with $\nu = \frac{1}{2}$. The reason for this choice of notation
will appear later. Similarly, one may calculate the rms radius of gyration, which also scales like $N^{1/2}$. The number of actual configu-
rations of a walk of N steps is simply μ^N, where $\mu = \int \delta(|\rho| - a) d^d \rho$
is a constant. As we shall see, in general this number behaves as
$N^{\gamma - 1} \mu^N$ for large N: this defines the exponent γ which, for simple
random walks, is unity. There is a strong element of universality
in these results for the exponents, since their values depend only
on the assumption that $\langle \rho_i{}^2 \rangle$ is finite and that there are no long
range correlations in the orientations of the monomers.

9.2 The Edwards model and the Flory formula

The above remark suggests that we may use instead a Gaussian
distribution for the vectors ρ_i, with a probability distribution pro-
portional to $\exp(-\rho_i{}^2/a^2)$. If we write $\rho_i = \mathbf{r}_{i+1} - \mathbf{r}_i$, where $\mathbf{r}_i$
represents the position of the ith vertex, then the steric repulsion
term may be modelled simply by adding a term

$$u a^d \sum_{i,j} \delta^{(d)}(\mathbf{r}_i - \mathbf{r}_j) \qquad (9.2)$$

to the hamiltonian, with u a dimensionless coupling constant. Finally, if we are interested in only the large scale properties of
the chain, we may formally take the continuum limit $a \to 0$ with
$\sum_i \to \int dt/a^2$ to obtain the *Edwards hamiltonian* for a single chain

$$\mathcal{H} = \int (dr/dt)^2 dt + u a^{d-4} \int \int \delta^{(d)}(\mathbf{r}(t_1) - \mathbf{r}(t_2)) dt_1 dt_2. \qquad (9.3)$$

Note that the scaling $t \sim a^2$ is necessary in order to make the coefficient of the $(dr/dt)^2$ term independent of a. This reflects the usual behaviour $r(t) \sim a(t/a^2)^{1/2}$ characteristic of a simple random walk when $u = 0$.

It is instructive to view (9.3) as the hamiltonian of a one-dimensional system, of finite length Na^2, with a d-dimensional degree of freedom $\mathbf{r}(t)$ and a long range interaction given by the second term. In this picture, a^2 plays the role of a short distance cut off in t. We may now apply a simple renormalization group argument along the lines of that in Section 5.4 for the continuous spin Ising model. The explicit dependence on a implies that there is a fixed point at $u = 0$, and that, for $d > 4$, the interaction u is irrelevant there. Thus we expect the exponents $\nu = \frac{1}{2}$ and $\gamma = 1$, characteristic of a simple random walk, to be also correct for the self-repelling walk, with finite u, when $d > 4$. For $d < 4$, however, u is relevant, and, by analogy with conventional critical behaviour, we might expect it to flow to some non-trivial fixed point u^*. At this fixed point, the field $\mathbf{r}(t)$ may be expected to pick up some non-trivial scaling dimension x, which means that it should rescale by a factor b^{-x} under a rescaling $a \to ba$ of the cut off a. Note that, at the non-interacting fixed point, $x = 0$, since $\mathbf{r} \sim t^{1/2}$ independent of a. The anomalous scaling dimension at the non-trivial fixed point may be estimated by the following argument. Under the rescaling, the two terms in the hamiltonian scale as b^{-2x} and $b^{(d-4)+dx}$ respectively. Non-trivial behaviour at the fixed point should require these to be the same, hence $x = (4-d)/(d+2)$. Thus $\mathbf{r}(t)$, which must have the form $af(t/a^2)$ by dimensional analysis, scales with t as $t^{(1+x)/2}$. Setting $t = N$, the end-to-end distance therefore scales as N^ν, where $\nu = (1 + x)/2$ or

$$\nu = \frac{3}{d+2}. \tag{9.4}$$

This result, first derived by Flory by a different but equivalent argument, is quite remarkable, since it purports to be an exact result for a non-mean field exponent. In fact, as we shall see in the next section when the problem is mapped onto that of a magnetic system, it is not exact. This is because it ignores the other interacting terms in the hamiltonian which are generated under renormalization. Nevertheless, for very deep mathematical reasons of supersymmetry, it turns out to be exact when $d = 2$. Since it is also correct as $d \to 4$, the Flory result is numerically quite accurate in three dimensions, where it gives $\nu = \frac{3}{5}$ as compared with the best current estimates of $\nu \approx 0.588$ which come from the magnetic analogy.

9.3 Mapping to the $O(n)$ model

The deepest insight into the problem of polymer statistics comes from a remarkable mapping to the $n \to 0$ limit of a magnetic system with n-component spins and $O(n)$ symmetry, first discovered by de Gennes. For definiteness, let us consider a lattice model of *self-avoiding walks*, whereby the polymer is modelled by random walks on a lattice such that no site is visited more than once. All such walks of the same length N are given the same weight. Consider an $O(n)$ model of the type discussed in Section 6.5, with spins $\mathbf{s}(r)$ at each site. They are normalised so that $\mathrm{Tr}\, s_a(r) s_b(r') = \delta_{ab} \delta_{rr'}$. Instead of the usual $O(n)$ generalisation of the Heisenberg model, we choose a hamiltonian

$$\mathcal{H} = -\sum_{r,r'} \ln\left(1 + x\mathbf{s}(r) \cdot \mathbf{s}(r')\right), \qquad (9.5)$$

where the sum is over nearest neighbour bonds, as usual. This hamiltonian is chosen so the partition function has the simple form

$$Z = \mathrm{Tr} \prod_{r,r'} \left(1 + x\mathbf{s}(r) \cdot \mathbf{s}(r')\right). \qquad (9.6)$$

Now suppose that we expand Z in powers of x. If the lattice has B nearest neighbour bonds in total, there will be 2^B terms in the expansion. Each term may be represented by a diagram drawn on the lattice. For each factor in (9.6), we may choose either the term 1 or the term $x\mathbf{s}(r) \cdot \mathbf{s}(r')$. In the latter case, draw in a line connecting r and r', and in the former case, omit it. Before taking the trace, then, a typical term is represented by a diagram like that illustrated in Figure 9.1. On taking the trace, any vertex with an odd number of lines running into it will be associated with an odd power of the spin variable at that site, and will therefore give zero. We are left with diagrams with an even number of lines running into each site. Ignoring for the time being those with four

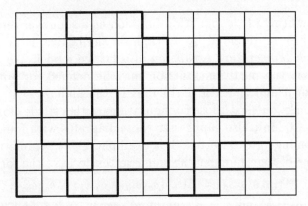

Figure 9.1. Diagram representing a typical term in the expansion of (9.6).

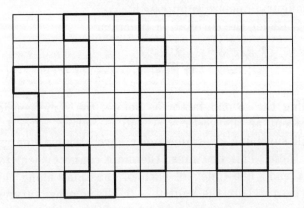

Figure 9.2. Typical diagram surviving after taking the trace in (9.6).

or more lines at a site, the only diagrams which survive will consist of closed, mutually self-avoiding loops, such as that illustrated in Figure 9.2. The trace will force the indices a of the spins around a given loop to be the same. Only on tracing over the last spin in each loop will a be unconstrained, giving rise to a factor of n for that loop. We conclude that the partition function may be rewritten as

$$Z = \sum_{\text{loop configurations}} n^{\text{number of loops}} x^{\text{number of bonds}}. \qquad (9.7)$$

This rewriting of the partition function as a sum over terms, each of which is represented by a graph on the lattice, is an example of a *high temperature expansion* of a lattice model. We see from (9.7) that the partition function may be defined for non-integer values of n and that, in particular, in the limit $n \to 0$ the term $O(n^1)$ picks out just those configurations with a single closed loop. The $n \to 0$ limit also suppresses those diagrams with four or more lines coming into a vertex. For example, $O(n)$ symmetry forces the trace of four spins at the same vertex to have the form

$$\text{Tr}(s_a s_b s_c s_d) = A(n)\left(\delta_{ab}\delta_{cd} + \delta_{ac}\delta_{bd} + \delta_{ad}\delta_{bc}\right). \qquad (9.8)$$

Setting $a = b$ and $c = d$, summing over a and c, and using the same normalisation as above, the left hand side gives n^2 while the right hand side is $n(n+2)A(n)$. Thus $A(n) \sim n/2$ as $n \to 0$ and such configurations are suppressed.

Now consider the correlation function

$$G(r_1 - r_2, x) \equiv \langle s_1(r_1)s_1(r_2)\rangle$$
$$= Z^{-1}\text{Tr}\prod_{r,r'} s_1(r_1)s_1(r_2)\left(1 + x\mathbf{s}(r)\cdot\mathbf{s}(r')\right). (9.9)$$

Expanding the second factor as before, the same considerations apply except at the vertices r_1 and r_2, where, after taking the trace, all surviving diagrams must have a single line going into each of these. This line must therefore connect them to avoid a dangling end. The index of each of the spins along this line is fixed to be $a = 1$. In the limit $n \to 0$, all the other surviving loops disappear, and we are therefore left with what we want: a sum over self-avoiding walks from r_1 to r_2, each weighted by a factor x raised to the power of its length N. Let us denote by $c_N(R)$ the number of self-avoiding walks between two points a distance R apart. Then the results of this section may be summarised by the relation

$$\sum_N c_N(R)x^N = \lim_{n\to 0} G(R, x) \qquad (9.10)$$

The connection between the two problems is therefore through a discrete version of a Laplace transform.

Critical behaviour

The above mapping was performed for a very specific hamiltonian on a lattice. However, in studying the critical behaviour, we may appeal to universality in two ways: first, we assert that the precise form of the hamiltonian (as long as the interactions are only short range) is not important, and the universal properties of the critical behaviour are therefore the same as the $n \to 0$ limit of the $O(n)$ generalisation of the Heisenberg magnet considered in Section 5.7; and that the choice of particular lattice is also unimportant. Why, though, is the *critical* behaviour of the $O(n)$ model relevant to the problem of polymer statistics? This is because the correlation function $G(R, x)$ is expected to be an analytic function of the temperature-like variable x with a finite radius of convergence. Since all the coefficients on the left hand side of (9.10) are non-negative, the closest singularity to the origin will occur on the positive x-axis, and will therefore correspond to the ferromagnetic critical point. The behaviour of $G(R, x)$ as x approaches this critical point is then dictated by the behaviour of the coefficients $c_N(R)$ at large N, which is what we wish to extract.

First, consider the total number $c_N \equiv \sum_R c_N(R)$ of N-step self-avoiding walks. Its generating function $\sum_N c_N x^N$ is, from (9.10), $\sum_R G(R, x)$, which is nothing but the susceptibility. This has a singularity as $x \to x_c-$ of the form $(x_c - x)^{-\gamma}$, which implies that, as $N \to \infty$,

$$c_N \sim \text{const}.N^{\gamma-1}\mu^N, \qquad (9.11)$$

where $\mu = x_c^{-1}$. The so-called *connective constant* μ, being related to the critical temperature, will be non-universal, and lattice dependent in particular, but the exponent γ is found by setting $n = 0$ in the results for the $O(n)$ model, and is universal. For $d > 4$ we thus find $\gamma = 1$, consistent with the result for a non-interacting random walk. In fact, the ordinary random walk corresponds to the Gaussian fixed point, as the self-avoiding walk corresponds to the Wilson–Fisher fixed point for $d < 4$.†

---

† It is not possible to use the results of the 2+ϵ expansion (see Section 6.5) since these apply only for $n > 2$. In fact, the $O(n)$ model with hamiltonian (9.5) has a non-trivial phase transition at a finite value of x in two dimensions when $n < 2$.

Similarly, the mean square end-to-end distance is given by

$$\langle R^2 \rangle = \sum_R R^2 c_N(R)/c_N. \tag{9.12}$$

The generating function for the numerator is $\sum_R R^2 G(R, x)$. The correlation function has, from Section 3.7, the scaling form

$$G(R, x) = \xi^{-(d-2+\eta)} \Phi(R/\xi) \tag{9.13}$$

near the critical point, so that $\sum_R R^2 G(R, x) \sim \xi^{4-\eta} \sim (x_c - x)^{-\gamma-2\nu}$, using the fact that $\xi \sim (x_c - x)^{-\nu}$ and the scaling relation $\gamma = \nu(2 - \eta)$ (see (3.53)). Thus

$$\langle R^2 \rangle \sim \text{const.} \frac{N^{2\nu+\gamma-1}\mu^N}{N^{\gamma-1}\mu^N} \sim \text{const.} N^{2\nu}, \tag{9.14}$$

justifying our use of the notation ν for this exponent earlier in this section. Putting $n = 0$ in the ϵ-expansion result (5.60) gives $\nu = \frac{1}{2} + \frac{\epsilon}{16} + O(\epsilon^2)$. Note that this does not agree with the Flory formula (9.4) in $4-\epsilon$ dimensions, which gives $\nu_F = \frac{1}{2} + \frac{\epsilon}{12} + O(\epsilon^2)$.

A more interesting quantity experimentally is the mean squared radius of gyration of self-avoiding walks of length N. For a single walk, this is given by $\langle R_G^2 \rangle = (1/2N^2)\sum_{r_1,r_2}(r_1 - r_2)^2$, where r_1 and r_2 are points visited by the walk. In order to count the walks which pass through two given points, we may consider the $n \to 0$ limit of the correlation function

$$\langle s_1(r_i)E(r_1)E(r_2)s_1(r_f) \rangle \tag{9.15}$$

in the $O(n)$ model. Here $E(r)$ is the energy density operator, defined on the bonds of the lattice as $\mathbf{s}(r) \cdot \mathbf{s}(r')$, where r and r' are neighbouring vertices. The same diagrammatic arguments as above show that this is proportional to the sum over all self-avoiding walks from r_i to r_f which contain the bonds at r_1 and r_2. Thus

$$\sum_{r_1,r_2,r_f} (r_1 - r_2)^2 \langle s_1(0)E(r_1)E(r_2)s_1(r_f) \rangle \propto \sum_N \langle R_G^2 \rangle_N c_N x^N, \tag{9.16}$$

where we have used translational invariance to set $r_i = 0$. Scaling now implies that the left hand side behaves as $\xi^{3d+2-2x_s-2x_E}$ near the critical point, where $x_s = \frac{1}{2}(d - 2 + \eta)$ and $x_E = d - 1/\nu$ are the scaling dimensions of the magnetisation and the energy density respectively. Substituting in these scaling relations, one then finds that $\langle R_G^2 \rangle \sim \text{const.} N^{2\nu}$, just as for the mean square end-to-end distance. In fact, it may be shown that the ratio of these two amplitudes depends only on universal quantities in the $O(n)$ model. It indicates that some aspects of the *shape* of large self-avoiding walks and linear polymers are also universal for $d < 4$.

9.4 Finite concentration

So far we have considered the statistics of a single polymer chain, which would be appropriate to the limiting case of a very dilute solution in which one may ignore the interactions between different chains. The case of a finite concentration is more difficult to analyse, but it is possible to make considerable progress using scaling arguments.† Suppose that the concentration of chains is $c = \rho_1/N$, where ρ_1 is the concentration of monomers, and N is the chain length (we assume that the polymers all have the same length, *i.e.* the solution is monodispersed.) Since we are discussing a solution, an appropriate quantity is the osmotic pressure $\pi(c, N)$, the pressure on a semi-permeable membrane which allows solvent but not the polymers to pass. In the limit $c \to 0$, this is given by the ideal gas law $\pi(c, N) \sim c k_B T$. It is therefore useful to define the *reduced* osmotic pressure $\tilde{\pi} \equiv \pi(c, N)/c k_B T$.

Now we may make a cross-over scaling argument. Suppose the microscopic length scale is changed by $a \to ba$ in such a way that the dimensionless physical quantity $\tilde{\pi}$ remains unaltered. The concentration will change by a factor b^d, but, in order to keep the physics the same, the average size R of an individual polymer should also change by a factor b^{-1}. This may be accomplished by changing N by a factor $b^{-1/\nu}$. We conclude that $\tilde{\pi}(c, N) = \tilde{\pi}(cb^d, Nb^{-1/\nu})$, or that

$$\tilde{\pi}(c, N) = Z(cN^{d\nu}/c_0) = Z(c/c^*), \qquad (9.17)$$

where Z is a scaling function and $c^* = c_0/N^{d\nu}$, with c_0 a constant. Expanding in powers of the polymer concentration c, we then find

$$\tilde{\pi} = 1 + B_2(N)c + O(c^2), \qquad (9.18)$$

---

† Although the parameter n of the $O(n)$ model corresponds to a fugacity for the chains, and therefore might be thought to describe a finite density solution when $n > 0$, in this ensemble the monomer fugacity x is the same for each chain. This therefore corresponds to an ensemble where the chains are free to exchange monomers.

whence the second virial coefficient $B_2(N)$ is predicted to behave as $N^{d\nu}$. The scaling form (9.17) predicts that, for finite c, once the single non-universal parameter c_0 is adjusted, the data from all different kinds of polymer solutions should fall on the same universal curve. This is indeed observed experimentally.

It is interesting to consider the case where N is large, with c fixed. In this case, the polymers will become increasingly entangled, and it becomes irrelevant where one ends and another begins. In this limit, we would expect that the properties of the system, in particular the osmotic pressure, should depend only on the concentration of *monomers* $\rho_1 = cN$. Thus

$$\pi = \frac{\rho_1 k_B T}{N} Z \left(\frac{\rho_1 N^{d\nu}}{c_0 N} \right) \tag{9.19}$$

should be independent of N in this limit. This implies that $Z(X) \sim X^{1/(d\nu-1)}$ as $X \to \infty$, so that, in this limit

$$\pi \sim \text{const.} \, \rho_1^{d\nu/(d\nu-1)}, \tag{9.20}$$

a remarkable prediction, since it depends only on the exponent ν defined for a single polymer.

9.5 Other applications

Many other concepts of equilibrium critical behaviour discussed in this book may be taken over to problems of polymer statistics. For example, the theta point, at which the effective interaction between two parts of the same chain changes sign from repulsive to attractive, corresponds, in mean field theory, to the vanishing of the M^4 term in the free energy of the $O(n)$ model, and therefore to a tricritical point. We know that the upper critical dimension for this problem is $d_c = 3$ (see Ex. 2.6), and therefore expect to find $\langle R^2 \rangle \sim N$ in this dimension, with, however, corrections going as a power of $\ln N$.

The subject of surface critical behaviour (Chapter 7) is also relevant. Modifying the exchange interaction in the $O(n)$ model near the boundary corresponds to a change in the effective monomer fugacity x in that region. This may occur if the monomers are attracted to the surface by some potential, when the phenomenon of *surface absorption* can occur. If this attraction is sufficiently weak, the entropic gain of executing a self-avoiding walk in the bulk dominates, and the polymer is not bound to the surface. This corresponds to the ordinary transition of the $O(n)$ model. The entropic exponent γ_s, which counts the number of self-avoiding walks with one end attached to the surface, is therefore different from that in the bulk. When the attraction is made sufficiently strong, the polymers are attached to the surface, and execute self-avoiding walks with $(d-1)$-dimensional exponents. This corresponds to the surface transition. The extraordinary transition of the $O(n)$ model really has no relevance for polymer statistics, since it is the closest singularity to $x = 0$, that is, the highest temperature transition, which dominates the large N behaviour. However, the above argument implies that it is the *special* transition which corresponds to the unbinding transition of the polymer from the surface.

As a final example, we consider the problem of *branched polymers*. These may be fabricated by sewing together long pieces of linear polymers. The statistics of these objects depend on whether the topology (the way in which the linear pieces are connected up) remains fixed as the total number of monomers N becomes large, or whether it becomes increasingly complicated. In the former case, a simple extension of the renormalization group approach to linear polymers suffices. In the mapping to the $O(n)$ model, each vertex where p linear polymers meet corresponds to the insertion of the particular scaling operator $\phi^{(p)} \equiv S_{a_1} \ldots S_{a_p}$, with all the a_j unequal. Suppose this has scaling dimension $x^{(p)}$. At the Gaussian fixed point, $x^{(p)} = px_{(1)} = p(d/2-1)$, but such a simple relation is no longer valid at the non-trivial fixed point. The generating function for polymers of a fixed topology, with vertices at $(r_1, r_2, \ldots)$ of respective degrees $(p_1, p_2, \ldots)$, is then proportional to the correlation function $\langle \phi^{(p_1)}(r_1)\phi^{(p_2)}(r_2) \ldots \rangle$. The total number of such polymers per unit volume has a generating function given by summing this over $(r_2, r_3, \ldots)$. This then scales like $\xi^{\sum_j(d-x^{(p_j)})-d}$, so that the number of such polymers behaves as $N^{\tilde\gamma-1}\mu^N$ where

$$\tilde\gamma = \nu\Big(\sum_j(d - x^{(p_j)}) - d\Big). \tag{9.21}$$

For the case of a linear polymer, this reduces to the previous relation (9.11) with $x^{(1)} = x_s = \frac{1}{2}(d-2+\eta)$. Relations like (9.21)

arise, therefore, as simple consequences of the existence of local
scaling operators with definite scaling dimensions.

When the topology grows increasingly complicated with N,
however, the problem is more difficult and turns out to correspond to a new universality class. The simplest ensemble to analyse in this case involves summing over all topologies, assigning,
in addition to a fugacity x for each monomer, a factor $u^{(p)}$ for
each vertex of degree p. On the lattice, the enumeration of such
self-avoiding branched polymer configurations of fixed large total
monomer number N is called the problem of *lattice animals*, and
numerical results suggest that all such models are in the same universality class, at least as long as the interactions are repulsive.
The corresponding continuous spin model may be written down
by analogy with linear self-avoiding walks. In that case, it was the
$n \to 0$ limit of the $O(n)$ model, with hamiltonian

$$\int \sum_a \left[((\nabla S_a)^2 + t S_a^2) + u \sum_{ab} S_a^2 S_b^2 \right] d^d r \qquad (9.22)$$

where the Gaussian terms may be thought of as corresponding
to ordinary random walks connecting the points at which the
self-repelling interaction u acts. To include branching vertices we
therefore simply add further interactions of the form $\sum_a \sum_p u^{(p)}$
$(S_a)^p$. For $p = 1$ this introduces a fugacity for free ends, for $p = 3$
for vertices of degree three, and so on. In order to analyse the
critical behaviour, it is first necessary to eliminate the linear term
in S_a by a shift $S_a \to S_a + \text{const.}$, and then, in sufficiently high
dimensions, to drop all but the first few powers of S since the rest
are irrelevant. This leaves an effective hamiltonian of the form

$$\mathcal{H} = \int \left[\sum_a ((\nabla S_a)^2 + t S_a^2 + u_3 S_a^3) + \Delta \sum_{ab} S_a S_b \right] d^d r, \qquad (9.23)$$

where the last term arises from the self-repelling interaction in
(9.22).†

This hamiltonian bears a striking resemblance to that for the
Ising model in a random field, (8.37). It may be thought of as
a model with an S^3, rather than an S^4, interaction, in a random field. For the reasons explained in Section 8.5, when $n = 0$

---

† There is also a term proportional to $\sum_{ab} S_a^2 S_b$, but it may be shown to be
irrelevant at the fixed point of interest.

the q^{-4} behaviour of the Fourier transformed correlation function in the Gaussian theory increases the upper critical dimension above that expected on the basis of simple dimensional analysis of the non-Gaussian term. We saw in the example of percolation (Section 8.4) that effective hamiltonians with cubic interactions generally have $d_c = 6$, but in the case of branched polymers, the anomalous scaling increases this to $d_c = 8$. As with the case of the random field Ising model, it is possible to carry out the perturbation expansion in $u_3^2 \Delta$, and to show that, to all orders the $8-\epsilon$ expansion of the random problem is identical to the $6-\epsilon$ expansion of the corresponding 'pure' problem. This turns out to be the Yang–Lee problem of the Ising model in a purely imaginary magnetic field.‡ This correspondence, called dimensional reduction, turns out not to be afflicted by the same non-perturbative effects which annul the corresponding argument for the random field Ising model (p.166), and we therefore conclude, for example, that the leading thermal exponent for the branched polymer problem in d dimensions is the same as that of the Yang–Lee problem in $d - 2$ dimensions. This remarkable result implies that $\nu = \frac{1}{2}$ for branched polymers in three dimensions, a result which is well confirmed by numerical studies. This is the only example of an interesting and non-trivial critical exponent in three dimensions whose value is known exactly.

Exercises

9.1 Calculate the mean square radius of gyration of a simple N-step random walk, and show that its N-dependence is controlled by the same exponent ν as the mean square end-to-end distance. Also show that this exponent controls the mean square radius of gyration of closed random walks, or loops.

9.2 Modify the argument leading to the Flory formula in Section 9.2 to the case of the theta point, by introducing a

‡ Since the sign of the $S_a S_b$ term is different in the two cases of the branched polymer problem and the random field Ising model, the former actually corresponds to theory with a real S^3 interaction in a purely imaginary random magnetic field. However, since the expansion parameter is $u_3^2 \Delta$, this is equivalent to a theory with a purely imaginary S^3 interaction, the Yang–Lee problem, in a random real field.

three-body interaction between the monomers, and setting the effective two-body interaction to zero.

9.3 Show that the number of *pairs* of mutually avoiding self-avoiding walks, which both begin and end at the same points, and whose *total* length is N, is simply related to the number of N-step self-avoiding polygons, or loops, whose asymptotic behaviour is given in terms of μ and ν. [Harder] What is the asymptotic behaviour of the number of pairs of such walks, each of which is now constrained to have the *same* length $N/2$?

9.4 The structure factor, $S(q)$, of a linear polymer, as measured, for example, in light scattering, is proportional to the Fourier transform of its density-density correlation function, conventionally normalised so that $S(0) = N$. Show that it is also related to an appropriate Fourier transform of the correlation function $\langle sEEs \rangle$ of the $O(n)$ model, discussed on p.176. Hence show that, in the limit $q \to 0$, $R \to \infty$ with qR fixed, it has the *universal* scaling form $S(q) = NF(qR)$. Work out the behaviour of $F(qR)$ for large values of its argument, and show that this is consistent with the scattering expected from an object of fractal dimension $1/\nu$.

9.5 Using the mapping to the $O(n)$ model, construct as many universal combinations of amplitudes related to the scaling behaviour of linear polymers as you can.

9.6 Using the methods of Section 3.5, calculate the $O(\epsilon)$ corrections to the scaling dimensions $x^{(p)}$ defined on p.179, and show that they are not simply proportional to p, as at the gaussian fixed point.

9.7 Suppose that the free energy of the model for branched polymers described by (9.23) has a singularity of the form $(t_c - t)^{2-\alpha}$, where t is the monomer fugacity. How does the number of such polymers of fixed total length N then behave for large N, assuming all the other parameters are held fixed? Show that the discussion of Section 8.4, together with the assumption that dimensional reduction holds, lead to the scaling law $\alpha = 2 - (d-2)\nu$, where the mean square radius of gyration behaves like $N^{2\nu}$.

10

Critical dynamics

So far, everything has concerned static, that is, time independent, phenomena, in equilibrium critical behaviour. The study of the *dynamics* of systems close to a critical point reveals an even greater richness of behaviour. For most of this chapter, we shall consider only examples where the departures from equilibrium are small. There is, in addition, a whole other underdeveloped field concerning systems which undergo phase transitions in some steady state far from equilibrium.

In considering the time dependence of systems close to equilibrium, there are two different kinds of observable quantities. For definiteness, consider a magnetic system. The first are the time-dependent generalisations of the correlation functions, for example, the two-point function

$$C(r - r', t - t') \equiv \langle s(r, t)s(r', t') \rangle, \tag{10.1}$$

whose Fourier transform may be measured by *inelastic* scattering. The expectation value is not, of course, calculable within the equilibrium Gibbs distribution, which is time independent. Rather, one should think of it as a long-time average

$$C(r - r', t) \equiv \lim_{T \to \infty} \frac{1}{2T} \int_{-T}^{T} s(r, t' + t)s(r', t')dt'. \tag{10.2}$$

The other important quantities are the *response functions*. These are defined by perturbing the system with a weak, time-dependent, external source (in this example, a magnetic field $h(r, t)$), and measuring the response of the system. For example, the simplest one-body response function satisfies

$$\langle s(r, t) \rangle = \int G(r - r', t - t')h(r', t')d^dr'dt'. \tag{10.3}$$

Note that causality implies that G vanishes when $t < t'$.

183

The correlation and response functions are related by the *fluctuation-dissipation relation*. This may be derived as follows. For clarity, we temporarily suppress the spatial dependence. In the absence of an applied field, a single measurement of the spin $s(t')$ must yield the values ± 1 with equal probabilities of $\frac{1}{2}$. Then, as in the time-independent case discussed in Section 2.3, we may write the correlation function as

$$\langle s(t)s(t')\rangle = \tfrac{1}{2}\langle s(t)\rangle_{s(t')=+1} - \tfrac{1}{2}\langle s(t)\rangle_{s(t')=-1}, \qquad (10.4)$$

where $\langle s(t)\rangle_{s(t')=+1}$ means the conditional expectation value of $s(t)$, summing over only those histories of the system in which $s(t') = 1$. Of course, in the absence of an external magnetic field the *sum* of the two terms on the right hand side of (10.4) vanishes by symmetry. Now imagine a scenario in which a small constant external field h is switched on at $t = -\infty$, and switched off again at time t'. In this case, the system will be in thermodynamic equilibrium up to this time, so that the probability that $s(t') = \pm 1$ is $e^{\pm h/k_B T}/2\cosh(h/k_B T) \approx \frac{1}{2}(1 \pm (h/k_B T))$. In this case, if we measure the magnetisation at time t, we shall find

$$\langle s(t)\rangle = \tfrac{1}{2}(1+(h/k_B T))\langle s(t)\rangle_{s(t')=+1}+\tfrac{1}{2}(1-(h/k_B T))\langle s(t)\rangle_{s(t')=-1}.$$
$$(10.5)$$

But, by the definition (10.3) of the response function, the $O(h)$ term in this expression is $\int_{-\infty}^{t'} G(t-t'')dt''$. Comparing with (10.4), it follows that, for $t > t'$,

$$C(t - t') = k_B T \int_{-\infty}^{t'} G(t - t'')dt''. \qquad (10.6)$$

This is valid for $t > t'$. The symmetry of $C(t - t')$ under time reversal then determines its behaviour for $t < t'$. (10.6) is the classical form of the fluctuation-dissipation relation.

It is more common to consider the Fourier transforms $\tilde{C}(r,\omega)$ and $\tilde{G}(r,\omega)$ of these functions with respect to time, because they have simple analytic properties as functions of ω. For example, causality implies that $\tilde{G}$ is analytic in the upper half plane. The fluctuation-dissipation relation (10.6) then reads

$$\tilde{C}(r,\omega) = \frac{2k_B T}{\omega}\mathrm{Im}\tilde{G}(r,\omega). \qquad (10.7)$$

Note that, in particular, the static limit $(t = t')$ of the correlation function is proportional to the zero frequency limit $\tilde{G}(\omega = 0)$ of the response function.

Away from a critical point, G usually decays exponentially with time like $e^{-t/\tau}$. As $T \to T_c$, the characteristic decay time τ diverges, in analogy with the correlation length. This is called critical slowing down. However, in systems with spontaneously broken continuous symmetry, the time dependence of the Goldstone modes typically leads to power law decays, even away from the critical point. It is important, therefore, to identify all these slow modes in a given physical situation.

In formulating models with which to describe critical dynamics, we are faced with a fundamental problem. Through the Heisenberg equations of motion, the microscopic quantum hamiltonian contains, in principle, all the required information about the time development of the density matrix. But these equations are time-reversal invariant, and it is not easy to see how they lead to the dissipative mechanisms characteristic of macroscopic systems at finite temperatures. Nor should we need to understand the details of this problem, since such systems are well described phenomenologically by a few important kinetic coefficients such as the viscosity, the heat conductivity, and so on. In practice, it is customary therefore to hide our ignorance about the detailed origin of irreversible dynamics, by writing down phenomenological equations of motion for only the coarse-grained degrees of freedom. These equations are constrained by two principles: first, the system, if left in isolation, should tend towards a Gibbs distribution; and, second, any conservation laws contained in the underlying microscopic dynamics should be retained in the coarse-grained picture. Within these broad constraints, one could of course write an infinitude of possible coarse-grained equations of motion. However, at least in the study of behaviour near a critical point, we are saved by universality: more or less any equation of motion should lead to the same universal properties, so we may as well choose the simplest possible description satisfying the above requirements. There are basically two approaches to writing down such equations. The first, based on a continuum picture, is more suitable for analytic work, while the second, a discrete approach, works better for simulations.

10.1 Continuum models

The prototype such equation, with only a single degree of freedom, describes the motion of a Brownian particle of unit mass in one dimension:

$$\dot{v}(t) = F(t) - \Gamma v(t) + \zeta(t). \tag{10.8}$$

Here $v(t)$ is the instantaneous velocity, and the terms on the right hand side represent the external force, the dissipative force, and the random noise due to collisions with the other microscopic particles. Note that the dissipative term may be written as $-\Gamma(\partial/\partial v)\mathcal{H}$, where $\mathcal{H}$, in this case, is just the kinetic energy $\frac{1}{2}v^2$. The noise satisfies $\langle \zeta(t)\rangle = 0$, and is supposed to be correlated significantly only over the time scales of the order of that between microscopic collisions. Over much longer time scales it is therefore permissible to assume that

$$\langle \zeta(t)\zeta(t')\rangle = 2D\delta(t - t'), \tag{10.9}$$

where D is a constant. Its value is determined by the requirement that, in the absence of the external force, the steady state distribution of v should be Maxwellian. Integrating (10.8) over a time interval δt gives

$$v(t + \delta t) \approx (1 - \Gamma\delta t)v(t) + \int_t^{t+\delta t} \zeta(t')dt'. \tag{10.10}$$

The two terms on the right hand side are uncorrelated, since $v(t)$ depends only on the values of the noise at earlier times. Hence

$$\langle v(t + \delta t)^2\rangle \approx (1 - 2\Gamma\delta t)\langle v(t)^2\rangle + 2D\delta t, \tag{10.11}$$

using (10.9) to calculate the variance of the second term. Since $\langle v^2\rangle = k_B T$ in equilibrium independently of t, it follows that $D = \Gamma k_B T$. This is the *Einstein relation*.

The equation of motion for a magnetic or fluid system, with an infinite number of degrees of freedom, follows in complete analogy. The hamiltonian is now the continuous spin form (called, in this context, the Landau–Ginzburg free energy functional)

$$\mathcal{H} = \int [\tfrac{1}{2}(\nabla S)^2 + \tilde{t}S^2 + uS^4]d^d r, \tag{10.12}$$

where we have used the symbol $\tilde{t}$ for the mean field reduced temperature, since t is now reserved for time. The time-dependent

Landau–Ginzburg (TDLG) model is chosen so that, in the absence of noise, the system will relax back towards a minimum of the free energy. Thus we write

$$\frac{\partial S}{\partial t} = -\Gamma \frac{\delta \mathcal{H}}{\delta S(r, t)} + \zeta(r, t) \qquad (10.13)$$

where

$$\frac{\delta \mathcal{H}}{\delta S(r, t)} = -\nabla^2 S + 2\tilde{t}S + 4uS^3. \qquad (10.14)$$

When a system has many degrees of freedom, the kinetic coefficients Γ in general form a matrix. In this case, its rows and columns are labelled by r and r'. Thus $\Gamma(\partial \mathcal{H}/\partial S(r, t))$ is shorthand for

$$\int \Gamma(r - r') \frac{\delta \mathcal{H}}{\delta S(r', t)} d^d r'. \qquad (10.15)$$

The Einstein relation now has the form

$$\langle \zeta(r, t)\zeta(r', t') \rangle = 2\Gamma(r - r')\delta(t - t'). \qquad (10.16)$$

Note that we are now working in reduced units where $k_B T = 1$.

Gaussian model

The simplest case to consider is when $u = 0$. This is appropriate to a discussion of the correlation functions outside the critical region (where it yields the Ornstein–Zernike result (2.19) in the static limit), and to the critical correlation functions when $d > 4$. In the context of the dynamics this approximation is known as the van Hove theory. The TDLG equation is now linear, so it may be solved by Fourier transform. Denoting the Fourier transform of $S(r)$ by S_k, and similarly for the other quantities, we have

$$\frac{\partial S_k}{\partial t} = -\Gamma_k(k^2 + \tilde{t})S_k + \zeta_k, \qquad (10.17)$$

where $\langle \zeta_k(t)\zeta_{k'}^*(t') \rangle = 2\Gamma_k \delta(k - k')\delta(t - t')$.

In this approximation, the 'modes' S_k for different wave numbers k are decoupled. Since we are interested in the large distance properties, only the small k behaviour of Γ_k is relevant. At this point, two distinct possibilities emerge. The underlying dynamics might be such as to conserve the total spin $\int S(r)d^d r$. This could happen, for example, if the total spin operator commutes with the quantum mechanical hamiltonian, as in a Heisenberg ferromagnet. Similarly, if we are thinking of the lattice gas picture of the Ising model as representing the critical point of a fluid, we should expect that the dynamics will conserve the total mass, which translates into the magnetisation. For both of these examples, $\partial S_{k=0}/\partial t = 0$, so that $\Gamma_{k=0}$ must vanish. Since $\Gamma(r - r')$ should be a short range function, this means that Γ_k is analytic, and we may take $\Gamma_k \sim \Gamma_0' k^2$, where Γ_0' is a constant. This model with a *conserved* order parameter is then known (for general values of u) as *model B*.

On the other hand, in the magnetic case the spins may interact with the crystalline fields in such a way that the total spin is not conserved. In that case, we may take $\Gamma_k \sim \Gamma_0$, a constant. This gives *model A*. Let us consider this case first, in the Gaussian approximation. Taking the expectation value of the TDLG equation,

$$\frac{\partial \langle S_k \rangle}{\partial t} = -\Gamma_0(k^2 + \xi_0^{-2})\langle S_k \rangle. \tag{10.18}$$

Each mode therefore has its own relaxation time $\tau_k = (\Gamma_0(k^2 + \xi^{-2}))^{-1}$, where $\xi \propto \tilde{t}^{-1/2}$ is the mean field correlation length. Note that $\tau_{k=0}$ diverges like ξ^z, where $z = 2$ in this case. More generally, this defines the *dynamic scaling exponent z*. The result for τ_k may be written in the scaling form $\tau_k = \xi^z f(k\xi)$. This is an example of a dynamic scaling form, which will turn out to have more general validity. By adding a time-dependent magnetic field $h_k \delta(t)$ to the Landau–Ginzburg hamiltonian (10.12), it is straightforward to solve for the response function in this approximation. The result is

$$\tilde{G}(k, \omega) = \frac{1}{-i\omega/\Gamma_0 + k^2 + \xi^{-2}}. \tag{10.19}$$

We see that indeed this reproduces the Ornstein–Zernike form in the static limit $\omega = 0$.

In the case of a conserved order parameter, on the other hand, the decay times are given by

$$\tau_k = \frac{1}{\Gamma_0' k^2(k^2 + \xi^{-2})}, \tag{10.20}$$

so that once again there is dynamic scaling, but now with $z = 4$. From this simple example we learn the important fact that two systems may have the same static critical exponents, but be in different dynamic universality classes.

10.2 Discrete models

Consider an Ising model with hamiltonian $\mathcal{H} = \sum_{i,j} J_{ij} s_i s_j$.† Since this is a classical hamiltonian, everything commutes, and it contains no information about the dynamics. Once again, it is necessary to put this in by hand, in the form of a *master equation*, which gives the time development of the whole probability distribution of the microstates. Let us label these states by $\{s\} = (s_1, s_2, \ldots)$, and consider the probability $P(\{s\}; t)$ for the system to be in a particular state at time t. The time evolution may be either discrete or continuous. The first case corresponds to what happens in a Monte Carlo simulation. At each tick of the clock, we choose a particular spin at random, and decide whether or not to flip it, based on a random number whose distribution is determined by the classical hamiltonian. The case of continuous dynamics is, however, easier to study analytically. The universal critical properties of each dynamical model should be the same, and should, moreover, agree with those obtained from the TDLG approach.

In the continuous case, we may suppose that, in a very short time interval δt, processes in which more than one spin is flipped may be ignored. Let $w_j(\{s\})\delta t$ be the probability that s_j is flipped during this interval, given that the state of the whole system is $\{s\}$. Then the master equation has the form

$$\frac{d}{dt} P(\{s\}; t) = -\sum_j w_j(s_j, \ldots) P(s_j, \ldots; t)$$
$$+ \sum_j w_j(-s_j, \ldots) P(-s_j, \ldots; t). \quad (10.21)$$

The way to understand this equation is to imagine that we have a large ensemble of similar systems, which begin from the same state, but all of which have slightly different dynamics, so that $P(\{s\}, t)$ tells us the relative number of systems in a given state. The two terms on the right hand side then represent respectively

---

† In this section the sites of the lattice will be labelled by $(i, j, \ldots)$ rather than $(r, r', \ldots)$, for convenience.

the rate of depopulation and population of a given state, by flipping the spin s_j. The rates w_j are to a large extent arbitrary. They should satisfy the locality requirement that w_j depends only on the states of the spins adjacent to the site j, and also that the steady state solution of the master equation (with the left hand side vanishing) is given by the Gibbs distribution $P \propto e^{-\mathcal{H}(\{s\})}$. This means that

$$\sum_j \left(w_j(s_j, \ldots)e^{-\mathcal{H}(s_j, \ldots)} - w_j(-s_j, \ldots)e^{-\mathcal{H}(-s_j, \ldots)} \right) = 0. \quad (10.22)$$

One way to ensure this is to stipulate that each term vanish separately. This is equivalent to requiring that equilibrium be achieved locally as well as globally, and is called the *principle of detailed balance*. It may be shown from this principle that the Gibbs distribution is then the unique asymptotic solution of the master equation as $t \to \infty$. Detailed balance may be written

$$\frac{w_j(s_j, \ldots)}{w_j(-s_j, \ldots)} = \frac{\exp(-h_j s_j)}{\exp(h_j s_j)}, \quad (10.23)$$

where $h_j = \sum_i J_{ij} s_i$ is the local molecular field. One simple way to satisfy both this and the requirement of locality is to choose

$$w_j(\{s\}) = \tfrac{1}{2}\Gamma(1 - s_j \tanh h_j), \quad (10.24)$$

where Γ is a rate, with the dimensions of inverse time. This defines the *Glauber model*. Note that in this case the order parameter is not conserved, so we expect its dynamic critical behaviour to be in the same universality class as model A. It is possible to write down a similar microscopic model in which the order parameter is locally conserved, for example by choosing the elementary process to be the exchange of two oppositely oriented adjacent spins. This gives *Kawasaki dynamics*, which are supposed to be in the same universality class as model B.

To study the dynamic scaling properties of the Glauber model further, let us consider the case of one dimension. Of course, the equilibrium short range Ising model has no phase transition in one dimension, but at sufficiently low temperatures the correlation length is large, and there is still a scaling region. If we take the case of simple nearest neighbour interaction, of reduced strength J, the possibilities for the downwards flip of a given spin are shown in Figure 10.1, along with their corresponding rates.

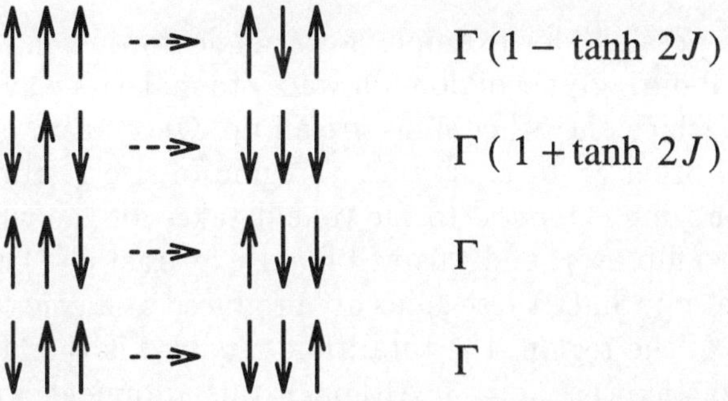

$$\uparrow\uparrow\uparrow \;\dashrightarrow\; \uparrow\downarrow\uparrow \qquad \Gamma\,(1 - \tanh 2J)$$

$$\downarrow\uparrow\downarrow \;\dashrightarrow\; \downarrow\downarrow\downarrow \qquad \Gamma\,(1 + \tanh 2J)$$

$$\uparrow\uparrow\downarrow \;\dashrightarrow\; \uparrow\downarrow\downarrow \qquad \Gamma$$

$$\downarrow\uparrow\uparrow \;\dashrightarrow\; \downarrow\downarrow\uparrow \qquad \Gamma$$

Figure 10.1. Elementary processes in the one-dimensional Glauber model.

At low temperatures, close to equilibrium, there will be large ordered regions separated by domain walls. The elementary processes illustrated in Figure 10.1 then correspond respectively to the creation and annihilation of pairs of walls, and to the motion of a single wall to the left or to the right. The last two processes do not alter the total number of walls. We may therefore write an equation for the time evolution of the density ρ of walls

$$\frac{d\rho}{dt} = -\Gamma(1 + \tanh 2J)\rho^2 + \Gamma(1 - \tanh 2J). \tag{10.25}$$

Thus ρ tends, at late times, to its steady state value

$$\rho^* = \left(\frac{1 - \tanh 2J}{1 + \tanh 2J}\right)^{1/2} = e^{-2J}. \tag{10.26}$$

This is precisely the inverse correlation length ξ^{-1} given by equilibrium statistical mechanics. At low temperatures, this density is small, and hence the dominant dynamical processes are the random walks of single domain walls. The relaxation time (the average time it takes for a given spin to flip) is then roughly the time taken for a single domain wall to move a correlation length, which is $O(\xi^2)$. Thus, for this model, the dynamic critical exponent is $z = 2$. The fact that this coincides with its value in the Gaussian model (presumably valid above four dimensions) might suggest that its value is independent of dimension. Actually, this is not the case.

In the case of conserved order parameter dynamics, isolated domain walls may not travel across large single phase regions. However, it is possible, for example, for a single down spin (equivalent to a neighbouring pair of domain walls) to make its way through a region in which the other spins are all up. Once again, it will take a time of order ξ^2 to cross such a region of size ξ. However, the relaxation time is related to the time it takes for the whole region of size ξ to flip over, and this will require at least $O(\xi)$ down spins to populate it. Since these spins are produced at a constant rate at the edge of the region, the total time required is therefore $O(\xi^3)$, so that, in this case, $z = 3$. Although this argument is somewhat heuristic, the result agrees with numerical simulations.

10.3 Dynamic scaling

Returning to the case of Glauber dynamics, let us see how these results might fit in with a simple renormalization group approach to the problem. The time it takes for a domain wall to move one lattice spacing is $O(\Gamma^{-1})$. Suppose now that we group the sites in blocks of size b. Since the domain wall is executing a random walk, the time it takes to cross a block will be $\Gamma^{-1}b^2$. The effect of the coarse graining is therefore to rescale the kinetic coefficient

$$\Gamma \to \Gamma' = b^{-z}\Gamma, \qquad (10.27)$$

where, in this case, $z = 2$.

This simple argument illustrates a general result of the renormalization group approach to dynamic critical behaviour. In addition to the usual renormalization group equations $\{K'\} = \mathcal{R}(\{K\})$ for the parameters $\{K\}$ defining the static hamiltonian, there is an additional equation expressing the renormalization of the rate Γ:

$$\Gamma' = \Gamma f(\{K\}). \qquad (10.28)$$

Note that this equation is linear in Γ, as it must be, since this is the only parameter in the model with the dimensions of inverse time. At the equilibrium critical fixed point then, $f(\{K^*\}) \equiv b^{-z}$ defines the dynamic exponent. More generally, a system may have several kinetic coefficients $(\Gamma_1, \Gamma_2, \ldots)$, and then the corresponding functions f_i may also depend on their ratios. Under renormalization, however, these ratios should themselves tend towards fixed point values (which may be finite, zero or infinity), and therefore it is still generally true that we may identify a single dynamic exponent z.

This picture gives an explanation of the phenomenon of dynamic scaling. Consider, for example, the response function $\widetilde{G}(r, \omega/\Gamma, \{K\})$. In writing this, we have used the fact that it can depend on the temporal quantities ω and Γ only through their ratio. In the static limit $\omega = 0$ the response function is the same as the static correlation function. In Section 3.7 we showed that this transforms under the renormalization group according to

$$\widetilde{G}(r, 0, \{K\}) = b^{2(y_h - d)} \widetilde{G}(r/b, 0, \{K'\}) \tag{10.29}$$

close to the fixed point. In terms of its Fourier transform with respect to r this reads

$$\widetilde{G}(k, 0, \{K\}) = b^{2-\eta} \widetilde{G}(bk, 0, \{K'\}), \tag{10.30}$$

and the generalisation to non-zero frequency is simply

$$\widetilde{G}(k, \omega/\Gamma, \{K\}) = b^{2-\eta} \widetilde{G}(bk, b^z \omega/\Gamma, \{K'\}). \tag{10.31}$$

Iterating in the usual way then gives the scaling form

$$\widetilde{G}(k, \omega, \xi) = \xi^{2-\eta} \Phi(\xi k, \xi^z \omega), \tag{10.32}$$

which is one of the ways of writing dynamic scaling for the response function. This means that the characteristic decay time for the decay of excitations with wave number k is $\tau_k = \xi^z f(k\xi)$. If we perform an inelastic scattering experiment at fixed angle (fixed k), then τ_k may be measured by fitting the peak in the dynamic structure factor to a Lorentzian form $1/(\omega^2 + \tau_k^{-2})$. In particular, for small angles, one should find that $\tau_0 \propto \xi^z$. At the critical point, ξ should disappear from the scaling formula (10.32), so that

$$\widetilde{G}(k, \omega) = k^{-2+\eta} \Psi(\omega k^{-z}). \tag{10.33}$$

The characteristic time is now $\tau_k \propto k^{-z}$.

10.4 Response functional formalism

The simplest way to understand critical dynamics is to formulate it so that it looks like a problem of equilibrium statistical mechanics in one extra dimension, much the same as was done for quantum critical behaviour in Section 4.5. This is achieved by the *response functional* formalism, which was first applied to the problem of fluid critical behaviour by Martin, Siggia and Rose. Consider for simplicity model A. The TDLG equation is

$$\dot{S} = -\Gamma(-\nabla^2 S + \tilde{t}S + uS^3 + h(r,t)) + \zeta(r,t), \qquad (10.34)$$

where we have added a source term $h(r,t)$. In calculating averages of quantities depending on S, we want to include only those values of S which satisfy the equation. We may ensure this by moving all terms to the left hand side, and writing this as the argument of a delta function:

$$\int [dS]\,\delta(\text{expression}) = \int [d\tilde{S}][dS]\,e^{-\int \tilde{S}(r,t)[\text{expression}]d^q r\,dt}.$$

$$(10.35)$$

We write a *functional* integral because the delta function constraint has to be applied for each value of r and t. The $\tilde{S}$ integral should, strictly speaking, lie along the imaginary axis.† It is now possible to average over the noise $\zeta(r,t)$. For each value of r and t this is a Gaussian integral

$$\int e^{\tilde{S}\zeta - \zeta^2/2\Gamma}d\zeta \propto e^{\Gamma\tilde{S}^2/2}. \qquad (10.36)$$

The result is that quantities like the correlation function $\langle S(r,t)S(r',t')\rangle$ may be evaluated as averages with respect to an ensemble with a 'probability distribution' proportional to $\exp(-\mathcal{H}(\tilde{S},S))$, where the 'hamiltonian' is

$$\mathcal{H} = \int \left[\tilde{S}(\dot{S} + \Gamma(-\nabla^2 S + tS + uS^3 + h(r,t))) - \tfrac{1}{2}\Gamma\tilde{S}^2 \right] d^q r\,dt. \qquad (10.37)$$

This looks rather like a continuous spin hamiltonian, with an extra dimension representing time, and a set of degrees of freedom $\tilde{S}$ in addition to the original S. In fact this is true only in a formal sense because $\mathcal{H}$ is not real, so that $e^{-\mathcal{H}}$ may not strictly be interpreted as a probability distribution. Nevertheless, the analogy is very useful in what follows. Note also that the time and space derivatives enter (10.37) in very different ways. Viewed as an equilibrium model, the theory is therefore expected to exhibit

---

† There is also, in principle, a Jacobian involved in writing (10.35) as an integral over S. However, if the left hand side of (10.34) is discretised as being proportional to $S(t + \delta t) - S(t)$, it is straightforward to show that this is unity. This is equivalent to defining the response function $G(r, t - t')$ to vanish when $t = t'$.

anisotropic scaling (see p.58). From the dynamical point of view, this is reflected in the fact that $z \neq 1$.

The variables $\widetilde{S}$ introduced above have a nice interpretation. The response function is obtained by differentiating the magnetisation $\langle S(r,t) \rangle$ with respect to the source $h(r',t')$ and then setting $h = 0$. Carrying this out explicitly with the weight $e^{-\mathcal{H}(\widetilde{S},S)}$, it follows that

$$G(r - r', t - t') = \langle S(r,t)\widetilde{S}(r',t') \rangle. \qquad (10.38)$$

For this reason, $\widetilde{S}$ is called the *response field*.

The advantage of this formalism is that we may apply all the machinery developed for the study of equilibrium critical phenomena. It turns out that the theory formulated in this manner remembers that it originated from a stochastic equation. This is encoded in a supersymmetry of h, which is responsible for the fluctuation-dissipation theorem being satisfied even in the presence of interactions.

It is easy to recover the results of Section 10.1 for the Gaussian model ($u = 0$) in this formalism. Although we have suppressed explicit powers of the microscopic length a, it is easy to restore them by dimensional analysis. There is a certain amount of arbitrariness in distributing the temporal rescaling between t and the rates Γ. We agree to assign no dimensions to time (consistent with the renormalization group scheme outlined above, where time was not rescaled). Then, denoting the dimension of a quantity X by $[X]$, we see from (10.37) that

$$[\widetilde{S}S] = k^2[\widetilde{S}\Gamma S] = [\Gamma \widetilde{S}^2] = k^d, \qquad (10.39)$$

where k denotes wave number. From this it follows that

$$[\Gamma] = k^{-2}, \qquad [\widetilde{S}] = k^{d/2+1}, \qquad [S] = k^{d/2-1}. \qquad (10.40)$$

The first result immediately implies that $z = 2$. From the others we may easily deduce that $[t] = k^2$ and $[u] = k^{d-4}$, thereby recovering the usual static results. As before, the upper critical dimension is $d_c = 4$. When $d = 4 - \epsilon$, it is possible to convert the results of an expansion in powers of u into an ϵ-expansion, as in Section 5.5. There will be an additional renormalization group equation describing the renormalization of Γ, which couples in (10.37) to the combination $\widetilde{S}^2$, for example. The $O(u)$ term in the renormalization group equation for Γ may thus be found by examining the operator product expansion of $\tilde{S}^2$ with $\tilde{S}S^3$. However, it is fairly easy to see that, using the Wick contractions $\langle S\tilde{S}\rangle$ and $\langle SS\rangle$ at our disposal, no $\tilde{S}^2$ terms gets generated in this product. Thus, there is no $O(u)$ correction to the renormalization of Γ, and, in fact, $z = 2 + O(\epsilon^2)$.

In the case of model B, with a conserved order parameter, the response functional formalism looks much the same, but with Γ replaced by the operator $-\Gamma'\nabla^2$. Dimensional analysis then shows that $z = 4$ at the Gaussian fixed point as expected, and, as above there is no $O(\epsilon)$ correction. However, in this case z turns out to be related to a static exponent. Consider the response function $G(r,t)$ Fourier transformed with respect to r. At the critical point, it has the scaling form

$$\tilde{G}(k,t) = k^{-2+\eta+z}\tilde{\Psi}(tk^z). \qquad (10.41)$$

From the TDLG equation for model B, however, the Fourier transform of S has an explicit factor of k^2 coming from the k-dependence of Γ_k. This will remain true even when $u \neq 0$. We therefore expect that $\tilde{G}(k,t) \sim k^2$ as $k \to 0$ at fixed t. A more careful examination of each term in the perturbation expansion in u confirms this. Comparing with the above scaling form, it follows that

$$z = 4 - \eta. \qquad (10.42)$$

This is consistent with the one-dimensional result $z = 3$ of Section 10.2, since the correlations in the Ising chain decay as r^0 for $a \ll r \ll \xi$, corresponding to $\eta + d - 2 = 0$, or $\eta = 1$.

10.5 Other dynamic universality classes

Effect of other diffusive modes

In studying equilibrium critical behaviour, it is permissible to restrict attention to only those degrees of freedom of the system, such as the order parameter fluctuations, which are critical. All other degrees of freedom, such as the phonons, are approximated by a heat bath in thermal contact with those which are treated explicitly. However, in discussing the dynamics, this is no longer adequate. We have seen that, according to the TDLG equation, the order parameter satisfies a diffusion-like equation at the critical point. But other degrees of freedom, for example the low wave number phonons which are responsible for the conduction of heat, also behave diffusively. In such circumstances the slow response of the heat bath may be the limiting relaxation process, and these modes should certainly be taken into account in the dynamical equations.

As an example, consider an Ising ferromagnet coupled to a phonon bath. The Landau–Ginzburg hamiltonian now depends on two sets of degrees of freedom: the order parameter S, and the local displacements of the atoms through the local energy density ρ of the phonons. The lowest order terms in the hamiltonian which respect the symmetry under spin reversal are

$$\mathcal{H} = \int [\tfrac{1}{2}(\nabla S)^2 + \tilde{t}S^2 + uS^4 + \tfrac{1}{2}w\rho^2 + g\rho S^2] d^d r, \qquad (10.43)$$

where, as we shall see below, w is proportional to the heat capacity of the phonons. Since we do not expect this to vanish at the critical point (unlike t), we need not include higher order terms such as $(\nabla \rho)^2$ or ρ^4, at least if we are interested in small wave numbers. The crucial term in (10.43) is the one coupling the energy density of the magnetic degrees of freedom to that of the bath.

If we are interested in only equilibrium properties, ρ may simply be integrated out of the partition function

$$\int e^{-\tfrac{1}{2}w\rho^2 - g\rho S^2} d\rho \propto e^{g^2 S^4 / 2w}, \qquad (10.44)$$

thereby merely shifting the value of u. For the dynamics, however, the story is different. The TDLG equations now take the form

$$\dot{S} = -\Gamma \frac{\partial \mathcal{H}}{\partial S} + \zeta \qquad (10.45)$$

$$\dot{\rho} = -D \frac{\partial \mathcal{H}}{\partial \rho} + \eta. \qquad (10.46)$$

Since the heat in the bath is conserved, D_k should behave as $D'k^2$ for small k. The second equation then becomes

$$\dot{\rho} = (D'w)\nabla^2(\rho + (g/2w)S^2) + \eta, \qquad (10.47)$$

which we recognise as the equation for heat conduction, with, however, the driving force being the gradient of the local energy density *including* that of the spin degrees of freedom.

The detailed analysis of the above model is too complicated to be included in this book. However, it is straightforward to establish a criterion for whether the inclusion of the effects of slow heat conduction will modify the dynamic universality class. Without loss of generality, we may set $w = 1$. The part of the response hamiltonian involving ρ is then

$$\int \left[\tilde{\rho}(\dot{\rho} - D'\nabla^2(\rho + \tfrac{1}{2}gS^2)) + D'\tilde{\rho}\nabla^2\tilde{\rho} \right] d^d r dt. \qquad (10.48)$$

The last term arises after averaging over the noise η, whose correlations must respect the conservation law obeyed by ρ. Dimensional analysis then gives $[D'] = k^2$ and $[\tilde{\rho}\rho] = [\tilde{\rho}^2] = k^d$, so that $[gk^{d/2}k^{d-2}] = k^d$, or $[g] = k^{2-d/2}$, where we have used the fact that $[S^2] = k^{d-2}$. This means that the coupling to the phonons is irrelevant for $d > 4$, but, at least at the Gaussian fixed point, it becomes relevant for $d < 4$, just as u does. However, we may do better. To determine whether g is relevant at the non-trivial fixed point for $d < 4$, simply insert the true scaling dimension k^{x_E} for the energy density S^2, where $x_E = d - 1/\nu$. We then find that

$$[g] = k^{(2-d\nu)/2\nu} = k^{\alpha/2\nu}. \qquad (10.49)$$

Thus the effects of slow heat conduction on the critical dynamics are irrelevant if $\alpha < 0$. For $\alpha > 0$, we expect a new dynamic critical behaviour with a different value of z.

The above argument was for model A. For model B it is straightforward to repeat the calculation, with the result that g is always irrelevant. This is because the relaxation of the spin degrees of freedom is so slow (like k^4 rather than k^2) that the heat bath has plenty of time to adjust itself.

Non-dissipative dynamics

In addition to damped, diffusive behaviour, many macroscopic systems are also capable of organised dynamical motion which results, for example, in the propagation of waves. How do such effects modify the dynamic critical behaviour? As a simple example, consider an isotropic Heisenberg ferromagnet, with order parameter **S** describing the local magnetisation. When any magnetic moment is subjected to an external **B**-field not aligned along its axis, it will precess according to the equation

$$\dot{\mathbf{S}} \propto \mathbf{S} \times \mathbf{B}, \tag{10.50}$$

where λ is a constant. In a Heisenberg ferromagnet, each local magnetic moment $\mathbf{S}(r)$ is subject to the molecular field, proportional to $\sum_{r'} J(r - r')\mathbf{S}(r')$, induced by the magnetisation of its neighbours. For small wave number excitations, this may be expanded, as usual, in the form $J(\mathbf{S}(r) + R^2 \nabla^2 \mathbf{S}(r) + \cdots)$. The cross product of the first term with $\mathbf{S}(r)$ vanishes, so that the leading term is proportional to $\mathbf{S} \times \nabla^2 \mathbf{S}$. This must be added to the usual model B dissipative terms. The result has the form

$$\dot{\mathbf{S}} = \lambda \mathbf{S} \times \nabla^2 \mathbf{S} + \Gamma' \nabla^4 \mathbf{S} + \cdots + \zeta, \tag{10.51}$$

where the neglected terms are the standard ones of model B. Note that the new term on the right hand side cannot be written as the functional derivative of some effective free energy. Equation (10.51) contains two quantities with the dimensions of inverse time, λ and Γ'. We need to ask whether the ratio λ/Γ' is relevant. Dimensional analysis at the Gaussian fixed point then gives

$$[\lambda/\Gamma'] = k^2[S]^{-1} = k^{3-d/2}, \tag{10.52}$$

so the new term is relevant when $d < 6$! This is surprising. It means that although $d = 4$ is the upper critical dimension for the statics, the dynamic critical behaviour will be non-Gaussian already for $4 < d < 6$. This may be investigated within a $6-d$-expansion, in which, at least for $d > 4$, it is permissible to take $u = 0$.

The above examples only partially illustrate the richness of possible behaviour in critical dynamics. For a given physical problem, it is crucial to identify all the possible slow modes, and to incorporate both their dissipative and non-dissipative dynamics in the equations. For fluids, for example, it is necessary to account for all the hydrodynamic modes. In somecases, this can lead to considerable complications. Indeed, apart from models A and B, a whole zoo of other TDLG-type models have been written down and analysed with renormalization group methods.

10.6 Directed percolation

The preceding sections described approaches to the dynamic behaviour of systems close to an equilibrium critical point. The analysis was guided by the principle that the correct description of the statics should appear in the appropriate limit. In writing down phenomenological equations, this dictates both the form of the dissipative terms, and of the noise, through the Einstein relations. The results of such analysis are then constrained by general relations such as the fluctuation-dissipation theorem. However, in describing critical behaviour of systems which are far from equilibrium, no such guiding principles are available. Indeed, it would appear that the whole subject should crumble into anarchy. However, once again, the renormalization group and universality provide tools for attempting a classification.

There is one very important dynamic universality class which is fundamentally different from those discussed so far, and is as ubiquitous in the study of non-equilibrium critical behaviour as is the Ising model in the equilibrium case. For reasons which will emerge, this has become known as the *directed percolation* universality class. Its simplest realisation, is however, in the realm of population dynamics.

Consider a population of individuals whose local density is described by the fluctuating variable $n(r, t)$. When n is small, there is a birth rate λ, but, when the local population gets too large, it is limited by a shortage of food. In the absence of fluctuations, such a system may be modelled by a rate equation of the form

$$\dot{n} = \lambda n - \mu n^2. \tag{10.53}$$

In this approximation, there is a dynamic phase transition at $\lambda = 0$. For $\lambda < 0$, the population always eventually dies out, while, for $\lambda > 0$, it achieves a steady state with density λ/μ. To describe the system more completely, however, the effects of diffusion and noise must be added. Thus (10.53) gets replaced by

$$\dot{n} = D\nabla^2 n + \lambda n - \mu n^2 + \zeta, \tag{10.54}$$

where D is the diffusion coefficient. In principle, this could depend on all powers of n and its derivatives, but all other terms will turn out to be irrelevant, at least near the upper critical dimension. By the same reasoning, we may neglect higher powers of n appearing on the right hand side. Since the steady states are not in thermal equilibrium, the correlations of the noise are not determined by the Einstein relation. However, they must satisfy the condition that, wherever $n = 0$, the noise vanishes also. This is because, in this example, the noise originates from the fluctuations in n itself, not from contact with a thermal bath. Thus the noise correlations themselves may be expanded in powers of n and its derivatives. The lowest order cumulants therefore have the form

$$\langle \zeta(r, t) \rangle = \lambda_1 n + D_1 \nabla^2 n - \mu_1 n^2, \quad (10.55)$$

$$\langle \zeta(r, t) \zeta(r', t') \rangle - \langle \zeta \rangle^2 = \mu' n \delta(r - r') \delta(t - t'), \quad (10.56)$$

where $(\lambda_1, \ldots, \mu')$ are constants, and all neglected terms will turn out to be irrelevant near the upper critical dimension. The terms in (10.55) may all be absorbed into a redefinition of the constants in (10.54) and will be henceforth ignored.

The critical behaviour may now be analysed using the response function formalism of Section 10.4, introducing a response field $\tilde{n}$. The average over the noise is now, however, given by a cumulant expansion

$$\left\langle e^{\int \tilde{n} \zeta \, d^d r \, dt} \right\rangle = e^{\int \mu' n \tilde{n}^2 \, d^d r \, dt + \cdots}, \quad (10.57)$$

so that the effective response hamiltonian, is, after an integration by parts

$$\mathcal{H} = \int [\tilde{n} \dot{n} + D \nabla \tilde{n} \cdot \nabla n - \lambda \tilde{n} n + \mu \tilde{n} n^2 - \mu' \tilde{n}^2 n] d^d r \, dt. \quad (10.58)$$

The dynamic phase transition which occurs in this model at $\lambda = \lambda_c$ (where now $\lambda_c \neq 0$ due to the fluctuations) is of a rather different nature from that in the dynamic Ising model. In the latter case, the transition is from a state in which $\langle S \rangle = 0$, but with fluctuations about this mean value, to one of two states with $\langle S \rangle \neq 0$, which break the $S \rightarrow -S$ symmetry of the model. In the example above, the state with $n = 0$ is always a possible steady state. For $\lambda < \lambda_c$, this is the only asymptotic state, while for $\lambda > \lambda_c$, there is another steady state, which is always reached unless $n(r, t) = 0$ initially. Moreover, there are no fluctuations about the trivial steady state with $n = 0$. For the Ising model, fluctuations always destabilise the $S = 0$ state when $T < T_c$. Nevertheless, it is straightforward to identify the order parameter of this transition as the value of $\langle n \rangle$ in the non-trivial steady state when $\lambda > \lambda_c$. The 'mean field' approximation, in this case, corresponds to the simple rate equation (10.53), from which we read off the mean field value $\beta = 1$ for the exponent describing the vanishing $(\lambda - \lambda_c)^\beta$ of the order parameter. The other classical exponents for this transition follow from the analysis of the Gaussian approximation with $\mu = \mu' = 0$. In this case the Fourier transform of the response function is

$$\tilde{G}(k, \omega) = (-i\omega + Dk^2 - \lambda)^{-1}, \tag{10.59}$$

so that, comparing with the standard notation of dynamic scaling in (10.32), we have the exponents $z = 2$, $\eta = 0$ and $\nu = \frac{1}{2}$. Here, ν describes the divergence $\xi \propto (\lambda - \lambda_c)^{-\nu}$ of the correlation length in the steady state when $\lambda > \lambda_c$. In the other phase, the correlation length cannot be defined because there are no fluctuations. However, the decay time $\tau \propto |\lambda_c - \lambda|^{-z\nu}$ of an initial fluctuation does make sense in either phase. Note that, while the response function $\langle n\tilde{n} \rangle$ is non-trivial, the correlation function $\langle n(r, t)n(r', t') \rangle$ vanishes for $\lambda < \lambda_c$. This is consistent with the lack of any fluctuation-dissipation theorem for this model.

The upper critical dimension follows from dimensional analysis on (10.58). Without loss of generality, we may rescale time and the response field so that $D = \mu' = 1$. Time then has dimensions (length)^2. We then have $[\tilde{n}n] = k^d$ and $[\tilde{n}^2 n] = k^{d+2}$, so that $[\tilde{n}] = k^2$ and $[n] = k^{d-2}$. Finally, this implies that

$$[\mu]k^2 k^{2d-4} = k^{d+2} \qquad \text{so that} \qquad [\mu] = k^{4-d}. \tag{10.60}$$

The non-Gaussian terms are therefore irrelevant at the Gaussian fixed point when $d > 4$, and the upper critical dimension of this problem is *four*. It is possible to calculate the exponents in an expansion in powers of $\epsilon = 4 - d$, although this is beyond the scope of this book. All the exponents ν, η and z gain non-trivial corrections at $O(\epsilon)$.

Finally, let us explain the connection with percolation. To do this, consider a discrete time version of the population dynamics discussed above. The simplest way to illustrate this, for $d = 1$, is on the diagonal square lattice illustrated in Figure 10.2. Time t is now supposed to flow up the page as indicated, and the one dimension of space is the horizontal axis. The density variable is replaced by an occupation number $n(r, t)$ for each site, which, for convenience, is restricted to take only the values 0 and 1. The dynamics is now described by a stochastic *cellular automaton*, that is, the value of

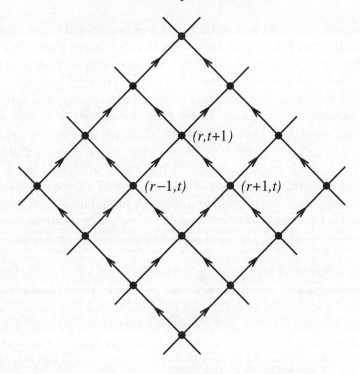

Figure 10.2. Directed percolation on the square lattice.

Table 10.1. *Rules for directed percolation.*

$n(r-1,t)$	$n(r+1,t)$	$P\big(n(r,t+1)=1\big)$
0	0	0
1	0	p
0	1	p
1	1	$2p-p^2$

$n(r,t+1)$ is determined only by the values of the $n(r',t)$ for values of r' close to r. On this particular lattice, one may choose it to depend only on $n(r-1,t)$ and $n(r+1,t)$. A simple set of rules is that shown in Table 10.1. In fact, these particular rules correspond to assigning a probability p for the *directed bonds* $((r-1,t) \to (r,t+1))$ and $((r+1,t) \to (r,t+1))$ to be independently present.

The site $n(r, t+1)$ is then defined to be occupied if it is connected by a directed bond to an occupied site at an earlier time. This then defines the problem of *directed bond percolation*. Oriented bonds are placed at random on the lattice, with probability p, according to the rule that their orientation should always be in the direction of increasing t. It is easy to see that the physics of this problem is very similar to that of the population model discussed earlier. If neither of the antecedent sites is occupied at time t, neither is $(r, t+1)$, with probability one. If just one is occupied, $(r, t+1)$ is occupied with probability p. However, if both antecedent sites are occupied, $(r, t+1)$ is occupied not with probability $2p$, but rather $2p - p^2$. This mimics the n^2 term in (10.53). The response function $G = \langle n(r, t)\tilde{n}(0, 0)\rangle$ is then proportional to the probability that (r, t) is occupied, given that, at 'time' $t = 0$, only the site $r = 0$ was occupied. This mapping is therefore yet another example of the mapping of a time dependent problem in d dimensions to a 'static' problem in $d+1$ dimensions. Note that, however, as is usual in such mappings, the static problem exhibits anisotropic scaling, characterised by an exponent $z \neq 1$. In that case, the exponent ν is sometimes denoted by $\nu_\perp$, and the divergence of the correlation length in the 'time' direction is characterised by $\nu_\parallel = z\nu_\perp$.

Exercises

10.1 Verify the fluctuation-dissipation relation for the Gaussian model described in Section 10.1.

10.2 Generalise the discussion of the fluctuation-dissipation relation on p.184 to the case of a non-zero constant magnetic field, and also, in this case, to higher order correlations like $\langle s(t)s(t')s(t'')\rangle$.

10.3 Show that if we try to modify the right hand side of the model A equation by including terms which may not be written in the form $\delta\mathcal{H}/\delta S$, but which are still consistent with the Ising symmetry and with rotational invariance, then such terms are irrelevant in $4 - \epsilon$ dimensions, in the sense of the renormalization group. Hence deduce that any such dynamical phase transition, whether or not it corresponds to an equilibrium system, should satisfy a form of the fluctuation-dissipation relation sufficiently close to the critical point.

10.4 On p.199 an example was given of a situation where the usual static coupling u is irrelevant at the dynamic fixed point. Assuming this remains true for $d < 4$, use the language of dangerous irrelevant variables (Section 3.6), to reconcile this with the known relevance of u for the static limit.

10.5 In the directed percolation problem, for $p > p_c$, the probability that the site (r, t) is connected to the origin is supposed to tend to some constant M for $|r| < Vt$, and to zero outside this expanding cone. Use scaling arguments to deduce the expected critical behaviour of the 'order parameter' M and the speed V as $p \to p_c$, in terms of the standard exponents $\nu_\perp$, η and z defined in the text.

10.6 A reaction-diffusion system in which particles A diffuse and undergo the reaction $A + A \to$ inert, at rate λ, whenever they meet, is described by the stochastic equation $\dot{n} = D\nabla^2 n - \lambda n^2 + \eta$, where, in contrast to directed percolation, the noise obeys $\langle \eta \rangle = 0$ and $\langle \eta(x, t)\eta(x', t') \rangle \propto R\delta(x - x')\delta(t - t')$, where $R \propto \lambda n^2$ is the reaction rate. Using the response function formalism, determine the upper critical dimension for this problem. Generalise to the case $kA \to$ inert.

11

Conformal symmetry

We saw in Chapter 3 that the hamiltonian for a system at a critical point flows under the renormalization group into a critical fixed point. Under a renormalization group transformation, the microscopic length scale is rescaled by a constant factor b, and so the coordinates of a given point, as measured in units of this length scale, transform according to $\mathbf{r} \to b^{-1}\mathbf{r}$. This is called a *scale transformation*. Once the flows reach such a fixed point, the parameters of the hamiltonian no longer change, and it is said to be *scale invariant*. As well as being scale invariant, the fixed point hamiltonian usually possesses other spatial symmetries. For example, if the underlying model is defined on a lattice, so that its hamiltonian is invariant under lattice translations, the corresponding critical fixed point hamiltonian is generally invariant under arbitrary uniform translations. This is because terms which might be added to the hamiltonian which break the symmetry under continuous translations down to its subgroup of lattice translations are irrelevant at such a fixed point. Similarly, if the lattice model is invariant under a sufficiently large subgroup of the rotation group (for example, if the interactions in the x and y directions on a square lattice are equal), then the fixed point hamiltonian enjoys full rotational invariance. As discussed on p.58, even if the interactions are anisotropic, rotational invariance may often be recovered by a suitable finite relative rescaling of the coordinates. For systems with intrinsic anisotropy, this is not the case, and we shall not discuss such cases further in this chapter. These operations of translation, rotation and scaling (dilatation) form a group. Under a general element of this group an arbitrary correlation function of scaling operators transforms in a simple way at the fixed point:

$$\langle \phi_1(r_1)\phi_2(r_2)\ldots\rangle = \prod_j b^{-x_j}\langle \phi_1(r_1')\phi_2(r_2')\ldots\rangle, \qquad (11.1)$$

206

where x_j is the scaling dimension of ϕ_j. In writing this, we have assumed that all the operators are scalars under rotation. Otherwise the appropriate rotation matrices need to appear on the right hand side.

It turns out, however, that as long as the fixed point hamiltonian contains only short range interactions, it is invariant under the larger symmetry of *conformal transformations*. For our purposes, a conformal transformation $r \rightarrow r'$ is one which *locally* corresponds to a combination of a translation, rotation and dilatation.† This is simpler to illustrate for the case of an infinitesimal transformation

$$r^\mu \rightarrow r'^\mu + \alpha^\mu(r), \tag{11.2}$$

where $\alpha^\mu(r) \ll 1$. If α^μ is a constant, this is of course simply a translation. When $\alpha^\mu(r)$ is slowly varying, the matrix of derivatives $\alpha^\mu{}_\nu = \partial \alpha^\mu / \partial r^\nu$ may be written as a sum of three pieces:‡

- an antisymmetric part $\alpha^{\mu,\nu} - \alpha^{\nu,\mu}$, which corresponds locally to a rotation;
- a diagonal part $\alpha^\lambda{}_{,\lambda} g^{\mu\nu}$, corresponding to a dilatation; and
- a traceless symmetric part $\alpha^{\mu,\nu} + \alpha^{\nu,\mu} - (2/d)\alpha^\lambda{}_{,\lambda} g^{\mu\nu}$, which may be thought of as the components of the local *shear*.

Conformal transformations are those for which this last piece vanishes. Since they have no shear component, they possess the property of preserving the angles between the tangents to curves meeting at a given point. An example of such a transformation in two dimensions is shown in Figure 11.1. The heuristic argument that invariance of the fixed point hamiltonian under translations, rotations and dilatations should imply its invariance under this larger set of symmetry transformations is deceptively simple. Imagine performing an inhomogeneous renormalization group transformation from the original regular lattice to one which is distorted by such a conformal mapping. In terms of a block spin transformation, this would mean replacing all the original degrees of freedom

---

† More correctly, conformal transformations should be viewed as acting on the *metric*, in such a way that $g_{\mu\nu} \rightarrow \Omega(r) g_{\mu\nu}$. This allows for conformal mappings between flat and curved spaces, for example. However, we shall restrict ourselves to flat spaces as these are more appropriate for statistical mechanics.

‡ Here, and throughout this chapter, we use the summation convention.

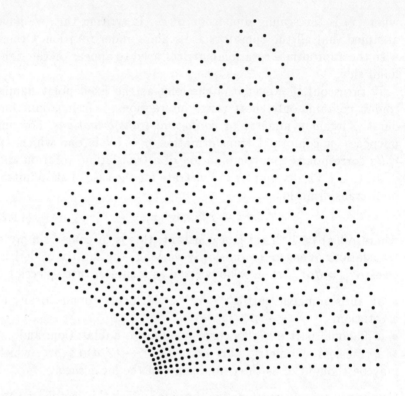

Figure 11.1. Example of a conformal transformation of part of a square lattice.

inside one cell of Figure 11.1 by a single block spin. In the vicinity of a given cell, the lattice spacing is rescaled by a factor $b(r)$, where $b(r)^{-d}$ is the Jacobian of the transformation $r \to r'$. If $b(r)$ is sufficiently slowly varying, then the way in which the local parameters of the hamiltonian transform in the neighbourhood of this cell will be just as if we were performing a uniform renormalization group transformation with rescaling factor $b(r)$ everywhere. Since the fixed point hamiltonian is invariant under such transformations, it is also invariant in the more general case when b varies with r. This argument can apply, of course, only if the fixed point hamiltonian is of sufficiently short range.

A simple generalisation of the reasoning in Section 3.7 which led to (11.1) now yields the transformation law for correlation functions

$$\langle \phi_1(r_1)\phi_2(r_2)\ldots\rangle = \prod_j b(r_j)^{-x_j}\langle \phi_1(r_1')\phi_2(r_2')\ldots\rangle, \qquad (11.3)$$

once again for scalar operators. Clearly the above argument is heuristic at best, and, in Section 11.3, we shall put it on a more systematic footing. In fact, it will be seen that (11.3) cannot be true for all scalar scaling operators. Indeed, it is a simple exercise to show that if it is true for the correlation functions of ϕ, it cannot be valid for its derivatives, for example $\nabla^2\phi$, even though these behave correctly under scale transformations. Instead, as we shall show, the transformation law (11.3) holds for a restricted class of scaling operators called *primary*. However, fortunately, the operators corresponding to the most relevant scaling variables are usually of this type.

11.1 Conformal transformations

The condition that an infinitesimal transformation be conformal is

$$\alpha^{\mu,\nu} + \alpha^{\nu,\mu} - (2/d)\alpha^{\lambda},_\lambda g^{\mu\nu} = 0, \qquad (11.4)$$

These equations are very restrictive when $d > 2$. In fact, the only solutions in that case, apart from infinitesimal translations, rotations and dilatations, are the so-called *special conformal transformations*

$$\alpha^\mu(r) = b^\mu r^2 - 2(b^\lambda r_\lambda)r^\mu. \qquad (11.5)$$

These may be thought of as made up of a finite conformal transformation, the *inversion* mapping $r^\mu \to r'^\mu = r^\mu/r^2$, followed by an infinitesimal translation by the vector b^μ, then a further inversion. Thus the conformal transformations for $d > 2$ may be generated by adding the discrete operation of inversion to the other three.

In two dimensions, however, there is far greater freedom. This may be seen most simply if we write (11.4) using *complex coordinates*, defined by $z = r^1 + ir^2$, $\bar{z} = r^1 - ir^2$. For most purposes, we may disregard the fact that $\bar{z}$ is the complex conjugate of z, and treat them as if they were independent complex variables. The line element in this coordinate system is

$$ds^2 = (dr^1)^2 + (dr^2)^2 = dz\,d\bar{z}, \qquad (11.6)$$

so that the metric is no longer diagonal. In fact

$$g_{\mu\nu} = \begin{pmatrix} 0 & \frac{1}{2} \\ \frac{1}{2} & 0 \end{pmatrix}. \tag{11.7}$$

For this reason, we must distinguish upper and lower indices. For example, for a vector $b_z = \frac{1}{2}b^{\bar{z}}$ and $b_{\bar{z}} = \frac{1}{2}b^z$. In these coordinates, the $(z\bar{z})$ and $(\bar{z}z)$ components of (11.4) are trivially satisfied for $d = 2$, and the others reduce to

$$\alpha^z{}_{,\bar{z}} = \alpha^{\bar{z}}{}_{,z} = 0. \tag{11.8}$$

Thus α^z depends only on z, rather than $\bar{z}$, which means that it is an *analytic* function of z. Similarly, $\alpha^{\bar{z}}$ is an analytic function of $\bar{z}$. This is the well known result that analytic functions correspond to conformal transformations in $d = 2$, and is the reason such mappings are so useful for solving Laplace's equation.

The notion of complex coordinates also makes it rather simple to discuss non-scalar operators in two dimensions. In general, we may classify such operators according to their *spin*.† Under a rotation $z \to ze^{i\theta}$, an operator of spin s transforms by a factor $e^{is\theta}$. What this means is that its two-point correlation function, for example, behaves like

$$\langle \phi(z_1, \bar{z}_1)\phi(z_2, \bar{z}_2) \rangle = |z_{12}|^{-2x}(\bar{z}_{12}/z_{12})^s, \tag{11.9}$$

where $z_{12} = z_1 - z_2$ and x is the usual scaling dimension of ϕ. Note that, for this two-point function to be single-valued, $2s$ should be an integer. (11.9) suggests that we define the so-called *complex scaling dimensions* $(h, \bar{h})$ by $x = h + \bar{h}$ and $s = h - \bar{h}$, so that the two-point function may be written $z_{12}^{-2h}\bar{z}_{12}^{-2\bar{h}}$. Note that $\bar{h}$ is *not* the complex conjugate of h. In fact, they are both real numbers. A simple consequence of this classification is that the operator product expansion of Section 5.1 has a simple form in $d = 2$, even for non-scalar operators:

$$\phi_i(z_1, \bar{z}_1) \cdot \phi_j(z_2, \bar{z}_2) = \sum_k c_{ijk} z_{12}^{-h_i-h_j+h_k} \bar{z}_{12}^{-\bar{h}_i-\bar{h}_j+\bar{h}_k} \phi_k(z_2, \bar{z}_2). \tag{11.10}$$

Similarly, the transformation law (11.3) for correlation functions under a conformal transformation corresponding to the analytic

---

† This has no physical connection with the quantum mechanical idea of spin.

mapping $z \to z' = w(z)$ may be written

$$\langle \phi_1(z_1, \bar{z}_1)\phi_2(z_2, \bar{z}_2) \ldots \rangle = \tag{11.11}$$

$$\prod_i w'(z_i)^{h_i} \overline{w'(z_i)}^{\bar{h}_i} \langle \phi_1(z_1', \bar{z}_1')\phi_2(z_2', \bar{z}_2') \ldots \rangle,$$

since the local dilatation factor is $|w'(z)|^{-1}$ and the local rotation is $\arg w'(z)$.

11.2 Simple consequences of conformal symmetry

In this section we shall assume the correctness of the transformation law (11.3) and deduce some simple consequences. The first set of results is valid for arbitrary dimension, since it exploits the symmetry under only special conformal transformations. For simplicity we then restrict the considerations to scalar operators.

Consider first the two-point correlation function of two different operators $\langle \phi_1(r_1)\phi_2(r_2) \rangle$. Conformal symmetry implies that this vanishes unless the scaling dimensions x_1 and x_2 are equal. The essence of the argument is simple. We can always choose a conformal transformation which maps the points r_1 and r_2 into, say, r_1' and r_2' respectively, under which the two-point function will be multiplied by a factor $b(r_1)^{-x_1}b(r_2)^{-x_2}$. Now imagine making the same transformation on $\langle \phi_2(r_1)\phi_1(r_2) \rangle$, where the two operators have been exchanged. This cannot affect the value of the correlation function, since they are related by a rotation through 180°. But now the rescaling factor will be $b(r_1)^{-x_2}b(r_2)^{-x_1}$. Since $b(r_1) \neq b(r_2)$ for a conformal mapping, the only way for these two results to agree when $x_1 \neq x_2$ is for the two-point function itself to vanish.

The above argument does not work when $x_1 = x_2$, so that if we consider the set of all operators ϕ_i with the same scaling dimension x, their two-point functions have the general form $\langle \phi_i(r_1)\phi_j(r_2) \rangle = d_{ij}r_{12}^{-2x}$. However, since d_{ij} must be real and symmetric, we may choose suitable linear combinations of the ϕ_i so that it is diagonal. In models satisfying *reflection positivity*,[†] (for

---

† This is true of most fixed points describing models with positive Boltzmann weights. Even microscopic models whose transfer matrix is not symmetric may correspond, at the fixed point, to a reflection positive theory. The main exceptions are cases like the $O(n)$ and Q-state Potts models for non-positive

example, when their transfer matrix may be brought into a symmetric form), these diagonal elements are all positive, and so, by normalising the operators appropriately, d_{ij} is simply δ_{ij}. Thus one of the simple consequences of conformal invariance is the *orthogonality* of scaling operators, in the sense that their two-point functions may be taken to have the form

$$\langle \phi_i(r_1)\phi_j(r_2) \rangle = \frac{\delta_{ij}}{r_{12}^{2x_i}}. \tag{11.12}$$

While such results sometimes also follow from the internal symmetries of the model (for example, the energy-magnetisation two-point function in the Ising model vanishes anyway on the grounds of the symmetry of the fixed point hamiltonian under reversing all the spins), we see that their provenance is more general.

For the three-point functions, conformal invariance completely fixes their functional dependence. To see this, note that, by translations, rotations and dilatations alone, two arbitrary points r_1 and r_2 may be mapped to two pre-assigned points. This is the reason why these symmetries are sufficient to fix the functional form of the two-point functions. The special conformal transformations then give one additional relation whereby *three* arbitrary points r_1, r_2 and r_3 may be mapped to three preassigned points r_1', r_2' and r_3'. Thus the three-point function $\langle \phi_i(r_1)\phi_j(r_2)\phi_k(r_3) \rangle$ may be related to the same correlation function with $r_i \to r_i'$, with the dependence on the r_i entering solely through the scaling factors $\prod_i b(r_i)^{x_i}$. The algebraic details of this calculation are not particularly illuminating, and it is simpler to verify the result, which has the remarkable elegant form

$$\frac{C_{ijk}}{|r_1 - r_2|^{x_i+x_j-x_k}|r_2 - r_3|^{x_j+x_k-x_i}|r_3 - r_1|^{x_k+x_i-x_j}}, \tag{11.13}$$

where C_{ijk} is a constant. In fact, this is equal to the operator product expansion coefficient c_{ijk} defined in Section 5.1, as long as the operators are correctly normalised as in (11.12) above. This follows immediately if we take the correlation function of both sides of the operator product expansion in (11.10) with the operator ϕ_k. Since we are free to permute the points in (11.13) without changing the value of the three-point function, it follows that the

integer n or Q, described in Sections 9.3 and 8.4.

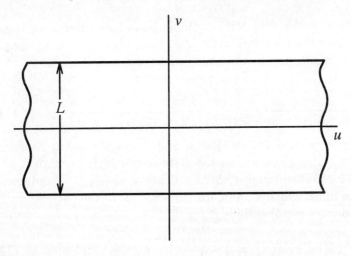

Figure 11.2. Finite-width strip geometry.

C_{ijk}, and therefore the operator product expansion coefficients (in the orthonormal operator basis) are totally symmetric functions of the indices (ijk). This is another very powerful result of conformal symmetry.

In two dimensions, conformal symmetry is much more powerful, because any analytic function $w(z)$ gives a conformal mapping $z \rightarrow z' = w(z)$. However, such a transformation will not, in general, map the plane onto itself, and so it is important to realise, in writing the transformation law (11.1), the correlation functions on either side may be evaluated in different geometries.† Sometimes this fact may be exploited, as when conformal mappings are used to transform a solution of Laplace's equation in one geometry to that in a simpler geometry.

Consider, for example, the mapping given by $w = (L/2\pi) \ln z$. This is analytic everywhere except at the origin, and maps the whole complex z-plane (minus this point) into the strip $|\text{Im} w| \le L/2$. A function which is single-valued in the plane will satisfy periodic boundary conditions between opposite edges of the strip (see Figure 11.2). This is an example of a quasi-one-dimensional finite size geometry discussed in Section 4.4. The form of the two-point

---

† The special conformal transformations do preserve the plane with the point at infinity added (the Riemann sphere).

function in the strip with periodic boundary conditions then follows from the transformation law (11.11) and its form in the plane, which, restricting to the scalar case for simplicity, is $|z_{12}|^{-2x}$. The result is

$$\frac{(2\pi/L)^{2x}}{[2\cosh{(2\pi(u_1 - u_2)/L)} - 2\cos{(2\pi(v_1 - v_2)/L)}]^x}, \tag{11.14}$$

where $w = u + iv$, so that u and v are Cartesian coordinates running along and across the strip respectively, as shown in Figure 11.2. When $|w_{12}| \ll L$, this behaves as $|w_{12}|^{-2x}$, independent of the finite width L, but, for $|u_1 - u_2| \gg L$, the correlation function decays exponentially

$$\langle \phi(u_1, v_1)\phi(u_2, v_2) \rangle \sim \left(\frac{2\pi}{L}\right)^{2x} e^{-(2\pi x/L)|u_1 - u_2|}. \tag{11.15}$$

Such an exponential decay is to be expected in a quasi-one-dimensional geometry. From (11.15) may be inferred the correlation length along the strip

$$\xi = L/(2\pi x). \tag{11.16}$$

The fact that the result is proportional to L is a consequence of finite-size scaling (see Section 4.4.) However, what is remarkable about (11.16) is that the amplitude ξ/L is simply related to the scaling dimension x. This prediction of conformal invariance in two dimensions has been amply verified by numerical and exact studies, and is now an important tool for extracting the scaling dimensions of otherwise unsolvable models. This is because the correlation length ξ is given in terms of the eigenvalue λ_j of the transfer matrix acting along the strip by the formula $\xi^{-1} = -\ln(\lambda_i/\lambda_0)$, where λ_0 is the largest, and λ_i is the dominant subleading eigenvalue which couples to the operator in question. For finite L, the transfer matrix is usually finite-dimensional, and is amenable to exact diagonalisation. In fact, since (11.15) is true for every (primary) operator, it follows that each scaling dimension x of the fixed point theory corresponds to an eigenstate of the transfer matrix. Actually, a stronger statement is true: there is a one-to-one correspondence between the full set of scaling operators at the fixed point, and the eigenvectors of the transfer matrix on the strip (at least those whose eigenvalues scale like L^{-1} as $L \to \infty$).

11.3 The stress tensor

A number of similar results follow from the transformation law (11.11), but, in order to understand better its theoretical under-pinnings, it is necessary to discuss the crucial role played by the *stress tensor*. This is a special scaling operator, which may be introduced as follows.† Suppose, instead of making a rescaling of the lattice which corresponds to a conformal transformation, that is, corresponds locally to a rotation and dilatation, we allow in addition the possibility of a shear component. If we imagine constructing a renormalization group transformation to this distorted lattice, there is no longer any reason to suppose that the fixed point hamiltonian will remain invariant. Instead, it will acquire an additional piece $\delta\mathcal{H}$, which, at least for an infinitesimal transformation, should be expressible as a linear combination of the complete set of scaling operators at the fixed point. If we consider the distortion of the lattice as corresponding to a general infinitesimal coordinate transformation $r^\mu \to r'^\mu + \alpha^\mu(r)$, the change in the hamiltonian may then be written

$$\delta\mathcal{H} = -S_d^{-1} \int T_{\mu\nu}(r)\partial^\mu\alpha^\nu(r)d^dr. \qquad (11.17)$$

This equation, for arbitrary $\alpha^\nu(r)$, then *defines* the stress tensor $T_{\mu\nu}$. The factor of S_d is the area of a unit hypersphere, and is conventionally introduced since it leads to greater convenience in later formulas. Once again, the crucial assumption of short range interactions has been invoked in writing (11.17), as it is implicitly assumed that $\delta\mathcal{H}$ may be written as a sum (integral) over local contributions, each of which depends only on the local distortion. Since, however, $\delta\mathcal{H}$ vanishes whenever this has components only of rotation and dilation, that is when the traceless symmetric part of $\partial^\mu\alpha^\nu$ vanishes, we see that $T_{\mu\nu}$ must be both symmetric and traceless itself.

It is important to realise that the stress tensor is a scaling operator just like the local energy density or magnetisation, although, unlike these, it is not a rotational scalar. As with these operators, it may be expressed as a linear combination of lattice operators expressed in terms of the fundamental degrees of freedom in the

---

† In quantum field theory, it is often known as the 'improved' energy-momentum tensor.

hamiltonian. For example, consider the critical Ising model on a square lattice, with equal nearest neighbour interactions K in the x and y directions. Suppose we make an infinitesimal shear transformation $x \to x' = (1 + \epsilon)x$, $y \to y' = (1 - \epsilon)y$, thus distorting the lattice. In terms of the new coordinates, the correlations will now be anisotropic. But we know, following the discussion on p.58, that this anisotropy may be accounted for by introducing anisotropic interactions $K_x \neq K_y$. In this case, we should take $K_x = K(1 - \lambda\epsilon)$ and $K_y = K(1 + \lambda\epsilon)$, where λ is some (non-universal) constant. This generates a new term proportional to a sum over $s(x, y)s(x + a, y) - s(x, y)s(x, y + a)$ in the hamiltonian. On the other hand, for this particular transformation, we see from (11.17) that $\delta\mathcal{H}$ is given by an integral over $T_{xx} - T_{yy}$. Hence, for this model,

$$T_{xx} - T_{yy} \propto s(x, y)s(x + a, y) - s(x, y)s(x, y + a)$$
$$+ s(x, y)s(x - a, y) - s(x, y)s(x, y - a), \text{(11.18)}$$

where we have antisymmetrised so that both sides have the same behaviour under reflections. It should be stressed that, in writing (11.18), as with all lattice identifications of scaling operators, it is valid only in the sense that correlation functions of either side are asymptotically the same when the points are far apart. At smaller separations, there are additional, less relevant, operators on the left hand side which will give rise to corrections.

From now on in this section, we shall consider only the case $d = 2$, where the use of complex components considerably simplifies the analysis. Why should one introduce a more general transformation if the aim is to analyse those which are conformal? The reason is that an analytic function $\alpha(z)$ cannot, in general, be small everywhere in the plane, but only in some finite region. Transformations which are infinitesimal everywhere must therefore be non-conformal at some points. Suppose we are interested in the effect of conformal transformations in the vicinity of the origin. Surround this by two regions $|z| < R_1$ and $R_1 < |z| < R_2$ (see Figure 11.3). Now make an infinitesimal transformation $r^\mu \to r'^\mu = r^\mu + \alpha^\mu(r)$ which is everywhere differentiable, and corresponds to the conformal transformation $z \to z' = z + \alpha(z)$ in the first region, while it reduces to the identity $r' = r$ for $|z| > R_2$. Otherwise, it is arbitrary in the annulus in between. The change

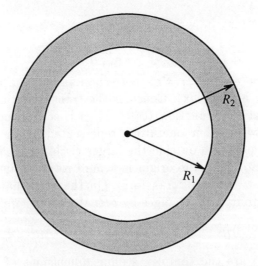

Figure 11.3. Geometry used to study the effect of a localised conformal transformation.

in the hamiltonian is, by (11.17), $\delta\mathcal{H} = -(1/2\pi)\int T_{\mu\nu}\partial^\mu\alpha^\nu d^2r$, where the integrand is non-vanishing only within the annulus. Integrating by parts then gives a term proportional to $\int \alpha^\nu \partial^\mu T_{\mu\nu} d^2r$, together with a surface term $\int \alpha^\nu T_{\mu\nu} dS^\mu$ on each circle. Since α^ν is quite arbitrary within the annulus, and since the result we are going to obtain from this calculation cannot depend on its precise form, then $\partial^\mu T_{\mu\nu}$ must vanish identically. Thus the stress tensor is also *conserved*. The surface term from $|z| = R_2$ vanishes, since $\alpha^\nu = 0$ there.

The stress tensor, being symmetric, has complex components $T_{zz} \equiv T$, $T_{\bar{z}\bar{z}} \equiv \overline{T}$ and $T_{z\bar{z}} = T_{\bar{z}z}$. In fact, the latter components vanish, since the trace is $T_z^z + T_{\bar{z}}^{\bar{z}} = 2T_{\bar{z}z} + 2T_{z\bar{z}}$. The conservation conditions then imply that

$$\partial^\mu T_{\mu z} = \partial^z T_{zz} = 2\partial_{\bar{z}} T = 0 \qquad (11.19)$$

$$\partial^\mu T_{\mu\bar{z}} = \partial^{\bar{z}} T_{\bar{z}\bar{z}} = 2\partial_z \overline{T} = 0. \qquad (11.20)$$

This means that correlation functions of $T(z, \bar{z})$ in fact depend only on z, so are therefore holomorphic functions. Similarly correlation functions of $\overline{T}$ are antiholomorphic.

In this notation, the boundary term in $\delta\mathcal{H}$ may be written as a contour integral around the circle $C : |z| = R_1$. After a little algebra, this may be written

$$\delta\mathcal{H} = \frac{1}{2\pi i}\int_C \alpha(z)T(z)dz - \frac{1}{2\pi i}\int_C \overline{\alpha(z)}\,\overline{T}(\bar{z})d\bar{z}. \qquad (11.21)$$

Now suppose that $\alpha(z) = \epsilon z$, corresponding to dilatation (and a rotation if ϵ is complex). Consider the transformation properties of the correlation function $\langle\phi(0)\dots\rangle$ of a scaling operator ϕ at the origin, with other operators, represented by the dots, all of whose arguments lie outside the larger circle $|z| = R_2$. Since the transformation near the origin is a pure rotation and dilatation, we know how $\phi(0)$ will transform. On the other hand, the fixed point hamiltonian $\mathcal{H}^*$ changes by $\delta\mathcal{H}$. We may therefore write

$$\langle\phi(0)\dots\rangle_{\mathcal{H}^*} = (1+\epsilon)^h(1+\bar{\epsilon})^{\bar{h}}\langle\phi(0)\dots\rangle_{\mathcal{H}^*+\delta\mathcal{H}}, \qquad (11.22)$$

where $(h,\bar{h})$ are the complex scaling dimensions of ϕ. The correlation function on the right hand side is to be evaluated with respect to the perturbed hamiltonian $\mathcal{H}^* + \delta\mathcal{H}$. This may also be written $\langle\phi(0)\dots e^{-\delta\mathcal{H}}\rangle_{\mathcal{H}^*}$. Expanding this exponential and equating coefficients of ϵ and $\bar{\epsilon}$ in (11.22), we then find

$$\frac{1}{2\pi i}\int_C zT(z)\phi(0)dz = h\phi(0), \qquad (11.23)$$

with a similar equation involving $\overline{T}$ and $\bar{h}$.

This has immediate consequences for the operator product expansion of $T(z)$ and $\phi(0)$. T itself has scaling dimensions $(2,0)$ since, from its definition (11.17), it has overall scaling dimension $x = 2$, and does not depend on $\bar{z}$, so $\bar{h} = 0$. The operator product expansion must be an analytic function of z, except possibly at $z = 0$, so may involve only integer powers of z:

$$T(z)\phi(0) = \sum_n z^{-2+n}\phi^{(n)}(0), \qquad (11.24)$$

thus defining the scaling operators $\phi^{(n)}$. Substituting this into the contour integral in (11.23) implies, by the residue theorem, that $\phi^{(0)} = h\phi$. An analogous argument, taking this time $\alpha = $ const., corresponding to a uniform translation, similarly gives $\phi^{(1)} = \partial_z\phi$.

Now note that the scaling dimension of $\phi^{(n)}$ is $h + n$. Since we do not expect to find scaling operators with arbitrarily negative dimensions, these $\phi^{(n)}$ must vanish for sufficiently large negative n. The special class of operators for which $\phi^{(n)} = 0$ for *all* $n < 0$

are called *primary*. For these, the leading terms in the operator product expansion with T are therefore already determined by their behaviour under translations, rotations and dilatations to be

$$T(z)\phi(0) = \frac{h}{z^2}\phi(0) + \frac{1}{z}\partial_z\phi(0) + \cdots. \qquad (11.25)$$

For such operators we now have a much stronger result. Consider an arbitrary infinitesimal conformal transformation, parametrised by $\alpha(z)$. By analogy with the above argument, the change in $\phi(0)$ is then given by the contour integral

$$\delta\phi(0) = \frac{1}{2\pi i}\int_C \alpha(z)T(z)\phi(0)dz - \frac{1}{2\pi i}\int_C \overline{\alpha(z)T}(\bar{z})\phi(0)d\bar{z}. \qquad (11.26)$$

Inserting the operator product expansion (11.25), the right hand side may be evaluated by Cauchy's theorem to give

$$\phi(0) \to \phi(\alpha(0),\bar{\alpha}(0)) + \left(h\alpha'(0) + \bar{h}\bar{\alpha}'(0)\right)\phi(0), \qquad (11.27)$$

that is, the transformation properties of a primary operator under conformal transformations are already determined by its behaviour under rotations and dilatations. This result is simple to generalise to the case of a finite conformal transformation $z \to z' = w(z)$:

$$\phi(z,\bar{z}) \to w'(z)^h \overline{w'(z)}^{\bar{h}} \phi(z',\bar{z}'), \qquad (11.28)$$

from which follows the transformation law for correlation functions (11.11) already exploited in the previous section.

Special conformal transformations, on the other hand, correspond to taking $\alpha(z) \propto z^2$, and the condition that an operator behave appropriately under this restricted class is that $\phi^{(-1)} = 0$. Such operators are called *quasi-primary*. Roughly speaking, such operators cannot be written as the derivatives of other operators with respect to the coordinates. These considerations extend, in fact, to dimensions $d > 2$. The results of Section 11.2 for the two- and three-point functions then rely only on the assumption that the operators involved are quasi-primary. Any scaling operator is either itself quasi-primary or may be written in terms of derivatives thereof.

It is important to realise that the stress tensor itself is not primary. In fact, its operator product expansion with itself must have the form

$$T(z)T(0) = \frac{c/2}{z^4} + \frac{2}{z^2}T(0) + \frac{1}{z}\partial_z T + \cdots. \tag{11.29}$$

The coefficient of z^{-2} term reflects the fact that T has $h = 2$. There is no z^{-3} term by symmetry under $z \to -z$. The z^{-4} term must be present so that, on taking the expectation value of both sides of the equation, the two-point function $\langle T(z)T(0) \rangle \propto 1/z^4$ is non-zero. Notice that, unlike other scaling operators, we are not free to adjust the normalisation of T so that its two-point function has coefficient unity. This is because this is already fixed by the definition of T through (11.17).

Equation (11.29) thus introduces what turns out to be a ubiquitous property of a fixed point theory in two dimensions, the so-called *conformal anomaly number*, or equivalently the *central charge*, c.† The additional term in (11.29) means that T does not transform in such a simple way as (11.28) under a finite conformal transformation. Instead, there is an additional term

$$T(z) \to w'(z)^2 T(z') - \frac{c}{12}\{z', z\}, \tag{11.30}$$

where $\{z', z\} \equiv (w'''w' - \frac{3}{2}w''^2)/w'^2$ is called the Schwartzian derivative.

Consider, for example, the mapping $w(z) = (L/2\pi)\ln z$ from the plane to the strip with periodic boundary conditions. In the plane, $\langle T \rangle = \langle \overline{T} \rangle = 0$ by rotational invariance. Equation (11.30) then yields the corresponding quantity in the strip

$$\langle T \rangle_{\text{strip}} = \langle \overline{T} \rangle_{\text{strip}} = \frac{c}{24}\left(\frac{2\pi}{L}\right)^2. \tag{11.31}$$

This gives the reduced free energy per unit length of the strip

$$E_0(L) = -\frac{1}{2\pi}\int_0^L \langle T_{uu} \rangle dv = -\frac{L}{2\pi}\left(\langle T \rangle_{\text{strip}} + \langle \overline{T} \rangle_{\text{strip}}\right). \tag{11.32}$$

To see this, recall the definition (11.17) of $T_{\mu\nu}$, and consider the transformation $(u, v) \to ((1 + \epsilon)u, v)$. The hamiltonian changes by an amount $\delta\mathcal{H} = -\frac{\epsilon}{2\pi}\int T_{uu}dudv$. The expectation value of this must be balanced by an explicit change in the free energy

---

† The first name originates from its role when the theory is defined on a curved background, when the trace T^μ_μ is non-zero and equal to $-cR/12$, where R is the scalar curvature. The second comes from the operator formulation of conformal symmetry, where it appears as the coefficient of the central term in the Virasoro algebra.

$-\epsilon E_0(L)$ per unit length. From (11.31), the free energy per unit length of the system in the strip geometry is therefore

$$E_0(L) = -\frac{\pi c}{6L}. \qquad (11.33)$$

Note that in writing this, we have already subtracted off an extensive $O(L)$ term proportional to the bulk free energy. Once again, the L-dependence of (11.33) is as expected from finite-size scaling, but the remarkable result is that its amplitude is related to the central charge of the fixed point theory. Since E_0 is simply given by the logarithm of the largest eigenvalue λ_0 of the transfer matrix, it may be extracted simply from numerical studies. This gives a direct way of measuring c.

A more physical interpretation of the central charge may be made if we consider the coordinate v across the strip as being imaginary time, and u as representing one-dimensional space. As discussed in Section 4.5, the partition function in this geometry may be viewed as that for a one-dimensional quantum system, at finite temperature $T = (k_B L)^{-1}$. Equation (11.33) then gives the reduced free energy per unit length $F/(k_B T)$ of this quantum system. From this we may read off the specific heat

$$C \sim \frac{\pi c k_B^2}{3} T. \qquad (11.34)$$

Since this result assumed rotational invariance of the equivalent two-dimensional model, it is valid only for systems with a dynamic exponent $z = 1$, that is, a linear dispersion law $\omega \sim v|k|$ for the elementary excitations. (11.34) is then written in units where $v = 1$. In addition, it is valid only when the width of the strip is much larger than the lattice spacing, which translates into low temperatures. With these provisos, we see that the low temperature behaviour of the specific heat of such a system is linear in T, and its slope is directly related to the value of c. It is instructive to compute the specific heat for free relativistic bosons in one dimension, using standard methods of statistical mechanics. This yields the form (11.34), with $c = 1$. In general, n types of noninteracting bosons would have $c = n$. The central charge may therefore be thought of as counting the number of gapless degrees of freedom of the theory. This interpretation must not be taken too literally, however, since interacting critical theories typically have non-integral values of c!

11.4 Further developments

Beyond this point, the study of conformal symmetry in two dimensions becomes increasingly mathematical and goes beyond the modest scope of this book. It is worth, however, recording some of the major results of the analysis.

As with any continuous symmetry in physics, the generators of conformal transformations form an algebra, called in this instance the Virasoro algebra. There are an infinite number of generators L_n, one for each term in the Laurent expansion of $\alpha(z) = \sum_n a_n z^{-n+1}$, and their commutator algebra has the form

$$[L_n, L_m] = (n - m)L_{n+m} + \frac{c}{12}n(n^2 - 1)\delta_{n,-m}. \tag{11.35}$$

In any particular fixed point theory, the L_n are represented by operators acting on the physical space of states. This is constructed by a technique known as radial quantisation. With this method, there is then a one-to-one correspondence between the scaling operators at the fixed point, and those states of the Hilbert space which are eigenstates of the operator L_0. This operator generates dilatations $\alpha(z) \propto z$, and its eigenvalues are just the scaling dimensions. The algebra allows one to organise the eigenstates of L_0, and hence the scaling operators, into irreducible representations of the algebra. The state in each representation with the lowest scaling dimension corresponds to a primary operator.

Many models of statistical mechanics satisfy the requirement of reflection positivity (see p.211). In this case, we are interested in unitary representations of the Virasoro algebra. It turns out that, for $c < 1$, these are severely constrained, in the same way that unitarity restricts the possible representations of a finite Lie algebra. In fact, only the values of c given by

$$c = 1 - \frac{6}{m(m + 1)}, \qquad \text{where } m = 3, 4, 5, \dots \tag{11.36}$$

are allowed, and, for each value of m, there are only a finite number of representations. The corresponding (complex) scaling dimensions of the primary operators are given by the Kac formula

$$h = h_{r,s} = \frac{(r(m + 1) - sm)^2 - 1}{4m(m + 1)}, \tag{11.37}$$

with $1 \leq s \leq r \leq m - 1$. This result goes at least part of the way towards realising the theorist's dream, implicit in the discussion of Chapter 4, of classifying *all* fixed points, and thereby all universality classes. Among the universality classes contained in the set with $c < 1$ are some old favourites: the critical Ising model, the 'hydrogen atom' of the subject, corresponds to $m = 3$, $c = \frac{1}{2}$. The tricritical Ising model sits at $m = 4$, the three-state Potts model at $m = 5$, and so on. For these models, it turns out that all universal properties of the fixed point theory can be found analytically: not only the scaling dimensions, but the operator product expansion coefficients, the correlation functions, and more.

11.5 The c-theorem

Recent work on the subject of two-dimensional critical models has focussed less on the conformally invariant fixed point theories themselves. Rather, it has attempted to elucidate the nature and universal properties of the renormalization group flows between them. A simple result is the so-called c-theorem of A. B. Zamolodchikov. Since it is relatively simple to state and prove, yet is deep and physically compelling, it forms a suitable point at which to end this account.

The c-theorem is formulated in the continuum limit. That is, it concerns the behaviour of renormalization group flows in the subspace of all interactions where the irrelevant lattice terms which break translational and rotational symmetry have already flowed to zero. Since rotational invariance will be a crucial input, this also excludes systems which exhibit intrinsically anisotropic scaling (see p.58). It also assumes reflection positivity (p.211). With these provisos, the c-theorem is simply stated:

• There exists a function C of the coupling constants which is nonincreasing along renormalization group flows, and is stationary only at the fixed points. Moreover, its value at each fixed point is that of the central charge c of the corresponding conformally invariant theory.

The proof relies on rotational invariance, reflection positivity, and the conservation of the stress tensor (a property which is a general consequence of translational invariance, and therefore is also valid away from the critical point). Consider some particular point on a renormalization group trajectory specified by a set of couplings $\{K\}$. For the time being, however, we suppress the dependence on $\{K\}$. Away from the fixed point, in addition to the components $T = T_{zz}$ and $\overline{T} = T_{\bar{z}\bar{z}}$ of the stress tensor, it has a non-zero trace $\Theta \equiv T_z^z + T_{\bar{z}}^{\bar{z}} = 4T_{z\bar{z}}$, since the hamiltonian is no longer invariant under dilatations. These three components, T, Θ and $\overline{T}$ have spins $s = 2, 0$ and -2 respectively under rotations. Thus their two-point functions have the form

$$\langle T(z, \bar{z})T(0,0)\rangle = F(z\bar{z})/z^4 \qquad (11.38)$$

$$\langle \Theta(z, \bar{z})T(0,0)\rangle = \langle T(z, \bar{z})\Theta(0,0)\rangle = G(z\bar{z})/z^3\bar{z} \qquad (11.39)$$

$$\langle \Theta(z, \bar{z})\Theta(0,0)\rangle = H(z\bar{z})/z^2\bar{z}^2 \qquad (11.40)$$

where F, G, and H are non-trivial scalar functions. On the other hand, conservation of the stress tensor $\partial^\mu T_{\mu z}$ implies, in complex coordinates, that

$$\partial_{\bar{z}}T + \tfrac{1}{4}\partial_z\Theta = 0. \qquad (11.41)$$

Taking the correlation function of the left hand side with $T(0,0)$ and $\Theta(0,0)$, respectively, yields two equations

$$\dot{F} + \tfrac{1}{4}(\dot{G} - 3G) = 0 \qquad (11.42)$$

$$\dot{G} - G + \tfrac{1}{4}(\dot{H} - 2H) = 0, \qquad (11.43)$$

where $\dot{F} \equiv z\bar{z}F'(z\bar{z})$, etc. On eliminating G and defining $C \equiv 2F - G - \tfrac{3}{8}H$, these reduce to

$$\dot{C} = -\tfrac{3}{4}H. \qquad (11.44)$$

Now reflection positivity requires that $\langle \Theta\Theta \rangle \geq 0$, so that $H \geq 0$. Thus C is a non-increasing function of $R \equiv (z\bar{z})^{1/2}$, and is stationary only when $H = 0$.

Now imagine making a renormalization group transformation $a \to a(1+\delta\ell)$. Since $C(R, \{K\})$ is dimensionless, this is equivalent to sending $R \to R(1 - \delta\ell)$, and the coupling constants $\{K\}$ will flow according to the renormalization group equations. Thus

$$\left(\frac{d}{d\ell} - R\frac{\partial}{\partial R}\right) C(R, \{K\}) = 0. \qquad (11.45)$$

If we now define $C(\{K\}) \equiv C(R_0, \{K\})$, where R_0 is some arbitrary but fixed length scale, we see that this quantity satisfies the first part of the c-theorem. Moreover, it is stationary if and only if $H = 0$, which, by reflection positivity, implies $\Theta = 0$, so that the theory is scale invariant and therefore corresponds to a fixed point. Finally, at such a fixed point, $G = H = 0$, and $F = \frac{1}{2}c$, so that indeed $C = c$.

The c-theorem has the interpretation that renormalization group flows go 'downhill'. In particular, it rules out the existence (for systems satisfying reflection positivity) of limit cycles and other esoteric behaviour in renormalization group flows. It also severely restricts the possible fixed points to which unstable directions at a given fixed point may flow. For example, relevant operators at the tricritical Ising fixed point, corresponding to $m = 4$ in the classification (11.36), may generate flows into either trivial fixed points with $c = 0$ (for example, high or low temperature fixed points), or to fixed points in the universality class of the critical Ising model, with $m = 3$, $c = \frac{1}{2}$. For this reason, the critical behaviour on the edges of the wings in the tricritical phase diagram shown in Figure 4.2 must be in the critical Ising universality class, despite the lack of any obvious symmetry.

An appealing physical interpretation of the C-function is as a kind of entropy of information about the critical system. Under renormalization, information is lost about the short distance behaviour of the correlation functions. However, this cannot be taken too literally – for example, even at infinite temperature a block spin transformation results in loss of information about the microstates of the system, yet no renormalization group flow takes place. Presumably, a more complete interpretation along these lines needs to account for the fact that the central charge is sensitive to only the effectively gapless degrees of freedom. Such a picture, if validated, would presumably extend to higher dimensions. However, so far, all attempts to prove higher-dimensional versions of the full c-theorem have failed. It is not difficult to satisfy the requirements of the theorem locally. The problem is to find a suitable function which is globally defined, is finite at each fixed point and is, at least in principle, measurable solely in terms of the correlation functions there.

Exercises

11.1 Show that the inversion transformation described in Section 11.1 is a conformal transformation, in any number of dimensions.

11.2 In the half space $z > 0$, when the order parameter is fixed to some non-zero value on the plane $z = 0$, its expectation value at the bulk critical point decays as z^{-x}, where x is its scaling dimension. By making an inversion about a suitable origin, find its behaviour in the interior of a sphere of radius R, with fixed boundary conditions on $r = R$.

11.3 By conformally mapping the upper half plane (with fixed boundary conditions on the order parameter) into a strip of width L, show that the correlation function along the strip decays exponentially at large distances, with a correlation length $\pi x^{(s)}/L$, where $x^{(s)}$ is the boundary scaling dimension of the order parameter (see Section 7.3).

11.4 By conformally mapping the upper half plane into a wedge of opening angle θ, show that an operator near the apex of the wedge has a scaling dimension $(\theta/\pi)x^{(s)}$, where $x^{(s)}$ is its boundary scaling dimension. Using scaling arguments, show that, below the bulk critical temperature the order parameter near the apex vanishes as $(-t)^{\beta(\theta)}$, and determine the dependence of this exponent on the opening angle.

11.5 The Gaussian model corresponding to the line of low-temperature fixed points of the two-dimensional XY model (Section 6.2) is a simple example of a conformally invariant system. Using the rules of Gaussian integration described in the Appendix, calculate the three-point function $\langle e^{iq_1\theta(r_1)}e^{iq_2\theta(r_2)} e^{iq_3\theta(r_3)}\rangle$ and show that it has the form given in (11.13). [Note that this correlation function vanishes unless $\sum_i q_i = 0$.]

11.6 Show that the specific heat of a one-dimensional gas of massless relativistic bosons has the form (11.34), and check that $c = 1$. Repeat the calculation for particles obeying Fermi-Dirac statistics. What is the value of c in this case?

Appendix: Gaussian integration

The simplest Gaussian integral over single real variable ϕ has the form

$$\int_{-\infty}^{\infty} d\phi \, e^{-\frac{1}{2}A\phi^2 + iJ\phi} = (2\pi/A)^{1/2} e^{-\frac{1}{2}A^{-1}J^2}. \qquad (A.1)$$

With respect to the Gaussian probability distribution proportional to $\exp\left(-\frac{1}{2}A\phi^2\right)$, then

$$\langle e^{iJ\phi} \rangle = e^{-\frac{1}{2}A^{-1}J^2}. \qquad (A.2)$$

This generalises to N variables $(\phi_1, \ldots)$ with a joint distribution proportional to $\exp\left(-\frac{1}{2}\sum_{rr'} \phi_r A_{rr'} \phi_{r'}\right)$:

$$\langle e^{i\sum_r J_r \phi_r} \rangle = e^{-\frac{1}{2}\sum_{rr'} J_r (A^{-1})_{rr'} J_{r'}}, \qquad (A.3)$$

and, finally, to the continuum case of a Gaussian distribution $\exp\left(-\frac{1}{2}\int\int \phi(r)A(r,r')\phi(r')d^dr\, d^dr'\right)$

$$W[J] \equiv \langle e^{i\int J(r)\phi(r)d^dr} \rangle = e^{-\frac{1}{2}\int\int J(r)A^{-1}(r,r')J(r')d^dr\, d^dr'}. \qquad (A.4)$$

The inverse A^{-1} is defined by

$$\int A^{-1}(r,r')\, A(r',r'')d^dr' = \delta^{(d)}(r - r''). \qquad (A.5)$$

From (A.4) follow several of the results used in the main text. For example, by differentiating with respect to $J(r_1)$ and $J(r_2)$, and setting $J = 0$, we find the two-point correlation function

$$\langle \phi(r_1)\phi(r_2) \rangle = A^{-1}(r_1, r_2). \qquad (A.6)$$

This means, in particular, that (A.4) may be rewritten

$$\langle e^{i\int J(r)\phi(r)d^dr} \rangle = e^{-\frac{1}{2}\int\int J(r)\langle\phi(r)\phi(r')\rangle J(r')d^dr\, d^dr'}. \qquad (A.7)$$

In most cases, $A(r,r')$ depends only on the difference $r-r'$, so that it is simpler to consider its Fourier transform $\tilde{A}(k) \equiv \int A(r-r')e^{ik\cdot r}d^dr$. In this basis, the matrix $\mathbf{A}$ is diagonal, so that it is simple to invert. As a result, (A.6) becomes

$$\langle \phi(r_1)\phi(r_2) \rangle = \int \tilde{A}(k)^{-1} \frac{d^dk}{(2\pi)^d}. \qquad (A.8)$$

227

The higher correlation functions all follow by taking suitable functional derivatives of $W[J]$ with respect to J, and then setting $J = 0$. The first time this happens, a factor $\int A^{-1}(r_1, r')J(r')d^d r'$ is brought down from the exponential. For this not to vanish on setting $J = 0$, $J(r')$ must be hit by another functional derivative $\delta/\delta J(r_i)$, leaving a factor of the two-point function $A^{-1}(r_1, r_i)$, with $i \neq 1$. Repeating this process then gives the version of Wick's theorem described on p.94.

Selected bibliography

Chapters 1–2   Stanley H.E., *Introduction to phase transitions and critical phenomena*. Oxford University Press, Oxford. (1971).
Yeomans J.M., *Statistical mechanics of phase transitions*. Clarendon Press, Oxford. (1992).
Fisher M.E., *Rep. Prog. Phys.* 30: 615. (1967).
Baxter R.J., *Exactly solved models in statistical mechanics*, Ch. 1–5. Academic Press, London. (1982).

Chapter 3   Fisher M.E., 'Scaling, universality and renormalisation group theory' in *Critical phenomena*, ed. Hahne F.J.W. Springer Lecture Notes, Berlin. (1983).
Pfeuty P. and Toulouse G., *Introduction to the renormalisation group and critical phenomena*. Wiley, London. (1977).
Goldenfeld N., *Lectures on phase transitions and the renormalisation group*. Addison–Wesley. (1992).
Niemeijer Th. and van Leeuwen J.M.J., 'Renormalisation in Ising-like spin systems' in *Phase transitions and critical phenomena*, vol. 6, eds. Domb C. and Green M.S. Academic Press, London. (1976).
Wegner F.J., 'The critical state, general aspects', *ibid.*

Section 3.8   Kadanoff L.P., 'Scaling at critical points', in *Phase transitions and critical phenomena*, vol. 5A, eds. Domb C. and Green M.S. Academic Press, London. (1976).

Section 3.9   Privman V., Hohenberg P.C. and Aharony A., 'Universal critical point amplitudes', in *Phase*

230   *Selected bibliography*

                     transitions and critical phenomena, vol. 14, eds. Domb C. and Lebowitz, J.L. Academic Press, London. (1991).

Section 4.1   Blume R., Emery V. and Griffiths R.B., *Phys. Rev. A* 4: 1071. (1971).

                     Lawrie I.D. and Sarbach S., 'Theory of tricritical points', in *Phase transitions and critical phenomena*, vol. 9, eds. Domb C. and Lebowitz J.L. Academic Press, London. (1984).

Section 4.2   Fisher M.E., *Rev. Mod. Phys.* 46: 597. (1974).

                     Aharony A., 'Universal critical behaviour', in *Phase transitions and critical phenomena*, vol. 6, eds. Domb C. and Green M.S. Academic Press, London. (1976).

Section 4.3   Sak J., *Phys. Rev. B* 15: 4344. (1977).

                     Brezin E., Le Guillou J.C. and Zinn-Justin J., *J. Phys. A*9: L119. (1976).

Section 4.4   Barber M.N., 'Finite-size scaling', in *Phase transitions and critical phenomena*, vol. 8, eds. Domb C. and Lebowitz J.L. Academic Press, London. (1983).

                     Cardy J.L., 'Finite-size scaling', in *Current Physics – Sources and Comments*, vol. 2. North–Holland, Amsterdam. (1988).

Section 4.5   Hertz J.A., *Phys. Rev. B* 14: 1165. (1976).

Section 5.1   Wilson K.G., *Phys. Rev.* 179: 1699. (1969).

Section 5.2   Anderson P.W. and Yuval G., *J. Phys. C* 4: 607. (1971).

                     Polyakov A.M., *Gauge fields and strings*. Harwood Academic Publishers. (1987).

Sections 5.3–5.5   Brezin E., Le Guillou J.C. and Zinn-Justin J., 'Field theoretical approach to the renormalisation group and its applications', in *Phase transitions and critical phenomena*, vol. 6, eds. Domb C. and Green M.S. Academic Press, London. (1976).

                     Wallace D.J., 'The ϵ-expansion and equation of state in isotropic systems', in *Phase transitions and critical phenomena*, vol. 6, eds. Domb C. and

Green M.S. Academic Press, London. (1976).

Amit D.J., *Field theory, the renormalisation group and critical phenomena* (2nd. edn.) World Scientific, Singapore. (1984).

Parisi G., *Statistical field theory*. Addison–Wesley. (1988).

Binney J.J., Dowrick N.J., Fisher A.J. and Newman M.E.J., *The modern theory of critical phenomena*. Clarendon Press, Oxford. (1992).

Le Guillou J.C. and Zinn-Justin J., *J. Physique Lett.* 46: L137. (1985).

Section 5.6   Wegner F.J., 'The critical state, general aspects', pp. 94–98, in *Phase transitions and critical phenomena*, vol. 6, eds. Domb C. and Green M.S. Academic Press, London. (1976).

Section 5.8   Aharony A., 'Universal critical behaviour', in *Phase transitions and critical phenomena*, vol. 6, eds. Domb C. and Green M.S. Academic Press, London. (1976).

Section 6.1   Mermin N.D. and Wagner H., *Phys. Rev. Lett.* 17: 1133. (1966).
Hohenberg P.C., *Phys. Rev.* 158: 383. (1967).
Coleman S., *Comm. Math. Phys.* 31: 259. (1973).

Section 6.2   Kosterlitz J.M. and Thouless D.J., 'Two-dimensional physics', in *Progress in low-temperature physics*, vol. VIIB. North–Holland, Amsterdam. (1978).
Berezinskii V.L., *Sov. Phys. JETP* 32: 493. (1970). *Sov. Phys. JETP* 34: 610. (1972).
Nelson, D. 'Defect-mediated phase transitions', in *Phase transitions and critical phenomena*, vol. 7, eds. Domb C. and Lebowitz J.L. Academic Press, London. (1983).

Section 6.4   Kosterlitz J.M., *J. Phys. C* 7: 1046. (1974).

Section 6.5   Amit D.J., *Field theory, the renormalisation group and critical phenomena* (2nd. edn.) World Scientific, Singapore. (1984).
Polyakov A.M., *Phys. Lett. B* 59: 79. (1975).

Chapter 7   Hohenberg P.C. and Binder K., *Phys. Rev. B* 9:

2194. (1974).

Binder K., 'Critical behaviour at surfaces', in *Phase transitions and critical phenomena*, vol. 8, eds. Domb C. and Lebowitz J.L. Academic Press, London. (1983).

Diehl H.W., 'Field-theoretic approach to critical behaviour at surfaces', in *Phase transitions and critical phenomena*, vol. 10, eds. Domb C. and Lebowitz J.L. Academic Press, London. (1986).

Section 8.1   Stinchcombe R.B., 'Dilute magnetism', in *Phase transitions and critical phenomena*, vol. 7, eds. Domb C. and Lebowitz J.L. Academic Press, London. (1983).

Section 8.2   Harris A.B., *J. Phys. C* 7: 1671. (1974).

Section 8.4   Stauffer D. and Aharony A., *Introduction to percolation theory* (2nd. ed.). Taylor and Francis, London. (1992).
Wu F.Y., *Rev. Mod. Phys.* 54: 235. (1982).

Section 8.5   Bray A.J. and Moore M.A., *J. Phys. C* 18: L927. (1985).

Chapter 9   de Gennes, P.G., *Scaling concepts in polymer physics*. Cornell University Press, Ithaca. (1980).

Section 9.2   Edwards S.F., *Proc. Phys. Soc.* 85: 613. (1965).
Flory P.J., *Principles of polymer chemistry*. Cornell University Press, Ithaca. (1953).

Section 9.3   de Gennes P.G., *Phys. Lett. A* 38: 339. (1972).
Nienhuis B., 'Coulomb gas formulation of two-dimensional phase transitions', in *Phase transitions and critical phenomena*, vol. 11, eds. Domb C. and Lebowitz J.L. Academic Press, London. (1987).

Section 9.4   des Cloizeaux J., *J. Physique* 36: 281. (1975).
Oono Y., *Adv. Chem. Phys.* 61: 301. (1984).

Section 9.5   Schäfer L., von Ferber C., Lehr U. and Duplantier B., *Nucl. Phys. B* 374: 473. (1992).
Parisi G. and Sourlas N., *Phys. Rev. Lett.* 43: 744. (1979).

Chapter 10   Hohenberg P.C. and Halperin B.I., *Rev. Mod. Phys.* 49: 435. (1977).
Ma S.-K., *Modern theory of critical phenomena.* Addison–Wesley. (1976).

Section 10.4   Martin P.C., Siggia E.D. and Rose H.H., *Phys. Rev.* A 8: 423. (1973).
Bausch R., Janssen H.K. and Wagner H., *Z. Phys.* B 24: 113. (1976).

Section 10.5   Halperin B.I., Hohenberg P.C. and Ma S.-K., *Phys. Rev. B* 10: 139. (1974).

Section 10.6   Duarte J.A.M.S., *Z. Phys. B* 80: 299. (1990).
Jensen I. and Dickman R., *J. Phys. A* 26: L151. (1993).

Chapter 11   Itzykson C. and Drouffe J.-M., *Statistical field theory*, vol. 2. Cambridge University Press, Cambridge. (1989).
Cardy J.L., 'Conformal invariance', in *Phase transitions and critical phenomena*, vol. 11, eds. Domb C. and Lebowitz J.L. Academic Press, London. (1987).
Cardy J.L., 'Conformal invariance and statistical mechanics', in *Fields, strings and critical phenomena*, eds. Brezin E. and Zinn-Justin J. North–Holland, Amsterdam. (1990).
Ginsparg P., 'Applied conformal field theory', *ibid.*
Christe Ph. and Henkel M., *Introduction to conformal invariance and its applications to critical phenomena.* Springer–Verlag, Berlin. (1993).

Section 11.5   Zamolodchikov A.B., *Pis'ma Zh. Eksp. Teor. Fiz.* 43: 565. (1986). [*JETP Lett.* 43: 730. (1986).]

Index

amplitudes, 55
  universal ratios, 7, 57
anisotropic scaling, 58, 80, 204
anisotropy
  spatial, 77
  uniaxial, 12, 67
annealed disorder, 145
anomalous dimension, 5
antiferromagnets, 8
asymptotic freedom, 131

bicritical point, 69
block spin transformation, 30, 37
Blume–Capel model, 62
boundary conditions
  fixed, 74, 137
  free, 74, 134
  periodic, 1, 74–76
Brownian motion, 186

c-theorem, 223
cellular automaton, 202
central charge, 220, 221
coarse graining, 30, 185
coexistence curve, 9
Coleman's theorem, 112
complex coordinates, 209
compressibility, isothermal, 10
confluent singularities, 48
conformal anomaly number, 220
conformal mapping, 213
conformal transformations, 207, 209
continuum limit, 53
corrections to scaling, 48
correlation functions
  critical exponents, 52
  density-density, 10
  dynamic, 183
  multipoint, 54
  Ornstein–Zernicke form, 24

orthogonality of two-point
  functions, 212
spin-spin, 4, 7
three-point functions, 212
transformation law, 51, 54, 207,
  209
correlation length, 1, 7
critical behaviour, 39
strip geometry, 214
transformation law, 37
critical dynamics
continuum models, 186
detailed balance, 190
directed percolation, 200
discrete models, 189
dynamic exponent, 8, 81, 188
dynamic scaling, 192
effect of other diffusive modes, 196
far from equilibrium, 200
Glauber model, 190
Kawasaki model, 190
model A, 188
model B, 188
non-dissipative terms, 198
renormalization group approach,
  192
response field, 195
response functional, 193
critical end point, 5
critical exponents, 3
branched polymers, 181
continuously varying, 115
directed percolation, 202
dynamic, 81, 188
ferromagnet, 7
from the renormalization group,
  45, 52
Gaussian fixed point, 92
hyperscaling relation, 52

234

mean field values, 20
percolation, 156
scaling relations, 46
self-avoiding walks, 175
simple fluid, 10
surface exponents, 136
critical isotherm, 10
critical surface, 41
cross-over behaviour, 61
cross-over exponent, 69
cross-over temperature, 70
due to quenched disorder, 150
in a random field, 162
long-range interactions, 71
multiple cross-overs, 72
percolation, 159
quantum to classical, 79
scaling function, 69
crystalline fields, 12
cubic symmetry breaking, 107
Curie temperature, 5

detailed balance, principle of, 190
dimensional reduction
branched polymers, 181
random field Ising model, 166
dimensional transmutation, 131
directed percolation, 200
critical exponents, 202
upper critical dimension, 202
discontinuity fixed points, 64
duality
Ising model, 151
XY model, 117
dynamic critical behaviour, 14

ϵ-expansion, 98
$O(n)$ model, 105
accuracy, 99
cubic symmetry breaking, 108
irrelevant operators, 100
Ising model, 94
surface critical behaviour, 142
Einstein relation, 186, 187, 200, 201
energy density, 54
near a boundary, 141
extrapolation length, 135

ferromagnet
n-component, 13

critical exponents, 7
dilute, 153
dynamic coupling to phonons, 196
Heisenberg, 8, 12
critical dynamics, 199
Ising, 12
uniaxial, 5, 12
uniaxial dipolar, 104
XY or planar, 12
Feynman path integral, 76
finite-size scaling, 72
amplitude-exponent relation, 214
free energy, 73
strip geometry, 213, 220
susceptibility, 73
fluctuation-dissipation relation, 184
fluctuations
corrections to mean field theory,
24
effects of, 4, 19, 113
in the random field Ising model,
166
longitudinal and transverse, 22
near a boundary, 133
thermal v. quantum, 12
fluids
binary, 14
simple, 8
critical exponents, 10
simple models, 13
free energy
finite-size scaling, 73
in strip geometry, 221
quenched average, 146
reduced, 44
scaling form, 45
singular part, 45
transformation law, 43

Gaussian integration, 227
Gaussian model, 92, 159
critical dynamics, 187
in random field, 165
in two dimensions, 114
surface critical behaviour, 140
Wick's theorem, 95
Gibbs distribution, 10, 185
Ginzburg criterion, 25, 94
Goldstone modes, 22, 113
Griffiths singularities, 150

Harris criterion, 148
Heisenberg model, 12
high-temperature expansions, 48, 99
hyperscaling, 52
  universal amplitude combination,
    57
  violation above upper critical
    dimension, 93
  violation in random fields, 164,
    166

Imry–Ma criterion, 162
irrelevant variables, 41, 48, 100
  dangerous, 49, 93, 103
Ising model, 12, 16, 30, 43
  continuous spin version, 90
  critical dynamics, 186, 189
  dilute, 146
    renormalization group
      approach, 150
  duality, 151
  existence of phase transition, 112
  in a random field, 161, 180
    dimensional reduction, 166
    Gaussian theory, 165
    line shape, 165
    lower critical dimension, 162
    renormalization group
      approach, 163
    upper critical dimension, 166
  in a transverse field, 77
  in one dimension, 34, 190
    lower critical dimension, 111
  multicritical behaviour, 66
  stress tensor, 216
  surface critical behaviour, 140
  upper critical dimension, 25, 92
  with vacancies, 61
isotherms, liquid–gas, 8

Josephson scaling, 132

Kac formula, 222
Kepler's law, 3
Kosterlitz–Thouless criterion, 117

Landau theory, 21
Landau–Ginzburg model, 92
  time-dependent, 187
Laplace's equation, 4, 136
lattice animals, 180

lattice gas, 13
Lifshitz point, 58
limit cycles, 225
long-range interactions, 71
lower critical dimension, 20
  continuous symmetries, 112
  discrete symmetries, 36, 111
  random field Ising model, 162

majority rule, 31
marginal variables, 41
  logarithmic corrections, 103
master equation, 189
mean field theory, 16
  continuous symmetries, 22
  correlation function, 22
  critical exponents, 20, 92
  fluctuation corrections, 24
  for surface critical behaviour, 134
  free energy, 18
  mean field equation, 18, 23
Mermin–Wagner–Hohenberg
    theorem, 112
models, relevance of, 11
molecular field, 18
Monte Carlo dynamics, 189
multicritical points, 66

$1/n$-expansion, 106
Néel temperature, 8
normal ordering, 95

$O(n)$ model, 13, 212
  $2 + \epsilon$ expansion, 129
  $n \to 0$ limit, 172
  large n limit, 105
  lower critical dimension, 112
  mean field theory, 22
  near four dimensions, 104
  near two dimensions, 127
operator product expansion, 84
  coefficients, 86
  in the Gaussian model, 95
  in two dimensions, 210
  with stress tensor, 219
order parameter, 6
Ornstein–Zernicke form, 24
osmotic pressure, 177

partition function, 10
Peierls argument, 112

percolation, 153
  bond, 155
  cluster size distribution, 157
  critical exponents, 156
  directed, 202
  effects of finite temperature, 159
  infinite cluster, 158
  mapping to the Potts model, 154
  mean cluster size, 156
  upper critical dimension, 158
phase transitions
  continuous, 2
  first-order, 2, 5, 20, 21, 62
    fluctuation-driven, 109
polymers
  branched, 179
    fixed topology, 179
    upper critical dimension, 181
  linear, 169
    at surfaces, 178
    critical exponents, 175
    Edwards model, 170
    finite concentration, 177
    Flory formula, 171
    mapping to $O(n)$ model, 172
    radius of gyration, 176
    random walk model, 169
    self-avoiding walk model, 172
    theta point, 178
    upper critical dimension, 171
population dynamics, 200
Potts model, 154, 212
  continuous spin version, 158
  critical behaviour, 156, 158
  upper critical dimension, 158

quantum effects, 12, 76
quantum electrodynamics, 28
quasi-long range order, 116
quenched disorder, 145
  random fields, 161
  self-averaging quantities, 147

random fields, 161
  cross-over behaviour, 162
  hyperscaling violation, 164, 166
reduced hamiltonian, 32
reduced variables, 7
  free energy, 44
redundant operators, 101

reflection positivity, 211, 223
relevant variables, 41
renormalization group
  beta functions, 47
  central idea, 29
  correlation functions, 49
  correlation length, 37
  critical dynamics, 192
  cross-over behaviour, 69
  discontinuity fixed points, 64
  eigenvalues, 41, 48
  equation, 35
  fixed points, 36, 40
  flows, 29, 36
  free energy, 43
  generation of further couplings, 39
  infinitesimal, 47
  linearised, 40
  near four dimensions, 98
  perturbative equations, 89
  real space, 30
  redundant operators, 101
  rescaling factor, 47
  scaling dimensions, 52
  scaling operators, 52
  scaling variables, 40
  surface critical behaviour, 138
  symmetries, 33, 42
  terminology, 28
  transformation, 33
  Wilson–Fisher fixed point, 94
replica method, 147, 165
replica symmetry breaking, 148, 167
response function, 183
rotational symmetry, 51, 206
roughening transition, 114, 117

scale factors, 45
scale invariance, 31, 206
scales, separation of, 4
scaling behaviour, 3
scaling dimensions, 52
  at a boundary, 136
  complex, 210
  Kac formula, 222
scaling functions, 45
scaling operators, 52
  at a boundary, 139
  primary, 209, 219
  quasi-primary, 219

scaling variables, 40
  at a boundary, 139
  magnetic, 42
  nonlinear, 56
  thermal, 42
self-averaging, 147
self-avoiding walks, 172
  connective constant, 175
  critical exponents, 175
self-similarity, 31
semi-infinite geometry, 133
solid-on-solid model, 117
specific heat
  critical behaviour, 7
  logarithmic corrections in four
    dimensions, 104
  one-dimensional quantum systems,
    221
spin waves, 113
spontaneous magnetisation
  critical behaviour, 7
stress tensor, 215
  in Ising model, 215
  operator product expansion, 219
  trace, 224
structural phase transitions, 14
superconductivity, 13, 25
superfluidity, 12, 114
  in two dimensions, 126
surface critical behaviour, 75, 133
  ϵ-expansion, 142
  surface exponents, 136
  boundary magnetisation, 136
  boundary scaling dimension, 136
  boundary scaling operators, 139
  boundary scaling variables, 139
  extraordinary transition, 137
  Gaussian fixed point, 140
  mean field theory, 134
  of polymers, 178
  ordinary transition, 134, 142
  renormalization group, 138
  special transition, 137, 141
  surface transition, 137
susceptibility, 4
  critical behaviour, 7
  finite-size scaling, 73
symmetry breaking
  cubic, 107
  spontaneous, 6

uniaxial, 67

transfer matrix, 77
translational symmetry, 206
tricritical behaviour, 178
tricritical point, 63
  effect of a magnetic field, 65
  renormalization group flows, 64
Trotter formula, 76

universality, 3, 10, 11, 14, 40, 43
  amplitude ratios, 57
  critical dynamics, 185, 189, 200
  critical exponents, 45
upper critical dimension, 26
  branched polymers, 181
  directed percolation, 202
  dynamic Heisenberg model, 199
  Ising model, 25, 92
  linear polymers, 171
  logarithmic corrections, 103
  percolation, 158
  polymer theta point, 178
  Potts model, 158
  random field Ising model, 166

Virasoro algebra, 222
vortices, 116, 122

Wick's theorem, 95, 228
Widom scaling, 46
Wilson–Fisher fixed point, 94

XY model, 12
  in two dimensions, 113
    critical behaviour, 124
    duality, 117
    Kosterlitz–Thouless criterion,
      117
    renormalization group
      equations, 122
    role of vortices, 116
    stiffness coefficient, 125

Yang–Lee problem, 181

Zamolodchikov's c-theorem, 223